MATHEMATICAL MUSIC

Mathematical Music offers a concise and easily accessible history of how mathematics was used to create music. The story presented in this short, engaging volume ranges from ratios in antiquity to random combinations in the 17th century, 20th-century statistics, and contemporary artificial intelligence.

This book provides a fascinating panorama of the gradual mechanization of thought processes involved in the creation of music. How did Baroque authors envision a composition system based on combinatorics? What was it like to create musical algorithms at the beginning of the 20th century, before the computer became a reality? And how does this all explain today's use of artificial intelligence and machine learning in music? In addition to discussing the history and the present state of mathematical music, Braguinski takes a look at what possibilities the near future of music AI might hold for listeners, musicians, and the society.

Grounded in research findings from musicology and the history of technology, and written for the non-specialist general audience, this book helps both student and professional readers to make sense of today's music AI by situating it in a continuous historical context.

Nikita Braguinski is a musicologist and historian of technology. He studied musicology at the University of Cologne and wrote his PhD in media theory at the Humboldt University of Berlin. He was a visiting postdoctoral fellow at the Max Planck Institute for the History of Science, a postdoctoral fellow of the Music Department at Harvard University, and most recently a postdoctoral researcher at Humboldt University where he wrote this book. He plays flute, piano, and guitar (but only when nobody is listening). His past musical experience ranges from playing in a rock band to jazz compositions, symphonic music, and electronic dance music.

MATHEMATICAL MUSIC

From Antiquity to Music AI

Nikita Braguinski

Routledge
Taylor & Francis Group

LONDON AND NEW YORK

Cover image: johnwoodcock / Getty Images

First published 2022
by Routledge
4 Park Square, Milton Park, Abingdon, Oxon OX14 4RN

and by Routledge
605 Third Avenue, New York, NY 10158

Routledge is an imprint of the Taylor & Francis Group, an informa business

British Library Cataloguing-in-Publication Data
A catalogue record for this book is available from the British Library

Library of Congress Cataloging-in-Publication Data
Names: Braguinski, Nikita, 1976– author.
Title: Mathematical music : from antiquity to music AI / Nikita
Braguinski.
Description: Abingdon, Oxon ; New York : Routledge, 2022. |
Includes bibliographical references and index. |
Identifiers: LCCN 2021043932 (print) | LCCN 2021043933 (ebook) |
ISBN 9781032062204 (hardback) | ISBN 9781032062198 (paperback) |
ISBN 9781003229254 (ebook)
Subjects: LCSH: Music—Mathematics. | Computer composition (Music) |
Artificial intelligence.
Classification: LCC ML3800 .B794 2022 (print) |
LCC ML3800 (ebook) | DDC 781.01/51—dc23
LC record available at https://lccn.loc.gov/2021043932
LC ebook record available at https://lccn.loc.gov/2021043933

ISBN: 978-1-032-06220-4 (hbk)
ISBN: 978-1-032-06219-8 (pbk)
ISBN: 978-1-003-22925-4 (ebk)

DOI: 10.4324/9781003229254

Typeset in Bembo
by codeMantra

CONTENTS

FIGURES

ACKNOWLEDGMENTS

Like every book, this one benefited greatly from the generous support of individuals and institutions. I would like to thank Sebastian Klotz for providing valuable comments for my initial idea behind this research project, and for hosting it at the music department of Humboldt University of Berlin. I am grateful for the Volkswagen Foundation's support of my research that made it possible for me to write this book. My thanks go to the organizers of the Sound Instruments and Sonic Cultures conference, to the Experimental Sound Research colloquium at the University of Halle-Wittenberg, and to Douglas Eck, who each gave me an opportunity to think through my argument by offering a place for a presentation and discussion. I wish to thank Alexander Rehding at Harvard University, Julia Kursell at Amsterdam University, and David Trippett at the University of Cambridge for their support of my research.

Funded by Volkswagen Foundation.

COMPOSING WITH NUMBERS (OVERVIEW OF THIS BOOK)

Music... Isn't it the perfect example of an art completely untouched by calculation? Just imagine yourself whistling in joy or singing a lullaby. It is not at all easy to believe that such activities are connected to math...

This book is an introduction to the centuries-long tradition of using mathematics to create music—how it started, where it stands today, and what the near future possibly holds for us. Primarily, it is about mathematics-inspired methods of *composition* such as the creation of melodies. As the book progresses through time periods, its focus shifts more and more towards popular genres, culminating in analyses of modern-day music generation and recommendation systems.

Today's use of artificial intelligence (AI) in music belongs to the larger area of mathematics-inspired methods of composition, and is by no means a singular, or 'revolutionary', development. In a sense, it has accumulated in itself traces of all the preceding mathematical ideas in music. Therefore, to learn about the history of mathematical music means also to learn about today's music AI, how it came to be, and where it may go.

Music AI is a fascinating, and for some, also frightening new topic. Is it going to prove the inferiority of human composers or, alternatively, automate away their jobs through mindless, but cheap imitations? The descriptions of historical, current, and developing approaches to automatic composition that are offered in this book provide the basis for a real understanding of the modern musician's situation.

DOI: 10.4324/9781003229254-1

Although the use of computers is indispensable for some of the musical-mathematical methods from the 20th and 21st centuries, this book is ultimately not about computers. Many people think that the use of AI to create music automatically is an outgrowth of computer technology, and the first half of the phrase (AI) certainly is. But the second half (automatic creation of music) is much older.

The purpose of this book is to enable you, the reader, to discuss critically the musical technologies of yesterday, today, and the near future, as well as the discourse that surrounds them. It fulfils this goal by embedding new technologies into a much longer history of what I call here mathematical music.

It offers easily understandable explanations of the core elements of the underlying mathematics and technology that make possible the acquisition of a deeper perspective without relying on formulas or technical details that are mostly only required for practical implementation.

It also combines ideas that come from different academic disciplines into one unified story. With this goal in mind, I avoid specialist terms, often replacing them with alternatives that are easier to understand, and explaining them when their use is really important for reasons of clarity.

As a whole, this book shows how the creative ideas about automatic composition accumulated over hundreds of years, rendering music AI thinkable and desirable, and it tries to imagine the scenarios that are possibly going to follow from today's situation.

Contents of this book

In the first part, I give a historical overview of the use of mathematics in music. Through this survey, I show how today's musical technologies are the heirs to previous mathematically inspired approaches, ranging from older uses of simple multiplication and division to the much more elaborate tools of the recent centuries. Starting with a short glimpse of prehistoric music technology such as bone flutes, the main section of this part discusses mathematical ideas about music coming from antiquity (ratios), the Middle Ages and the early modern period (combinatorics), the 19th century (acoustics), and the 20th century (statistics, algorithms, computers).

The second part is a discussion of recent and future developments in automatic and assisted composition. I explain how these developments are primarily grounded in the branch of artificial intelligence called machine learning and, specifically, deep learning. One of the chapters offers a non-technical explanation of what deep learning can realistically achieve in music today. Others show the possibility of mass-produced, but still individually tailored cultural products based on data, and argue for the importance of early 20th century avant-garde for an understanding of the future of pop.

The conclusion sums up some of the possible future developments in music AI. It takes a look at both the technological ideas (such as the imitation of typical limits of human physiology and cognition) and the societal issues (such as the misuse of music AI to create psychological profiles of listeners).

In writing this book, it was my goal to always explain any less common terms in the text itself. Additionally, I have collected most of these terms in the glossary at the end of the volume to facilitate later reference.

Thirteen chapters each focusing a special moment or subject together form a panorama of the gradual mechanization of thought processes involved in the creation of music. Some of the questions they answer are: What does today's music technology have to do with Pythagoreanism? How did Baroque authors envision a composition system based on combinatorics? How did the developing science of sound change the understanding of music in the 19th century? What was it like to create musical algorithms at the beginning of the 20th century, before the computer became a reality? And how does this all explain today's use of AI and machine learning in music?

As this book progresses from the past and our present moment to imaginations of the future, it necessarily turns from description to (informed) speculation. It goes without saying that some of the imagined future developments will not take place. But it is important to think about possibilities now, so we have a chance to have an opinion and a plan in case they do happen.

One of the core arguments of this book is that a common history unites the earliest theories from antiquity with today's musical AI. Combining insights coming from musicology and the history of technology, this overview enables the non-specialist to get rid of popular myths surrounding music and AI as it "translates" research for the interested reader.

Businesses built on the use of AI in music are developing at a high pace, and their attempts to present their technologies as a "breakthrough" or a "revolution" maybe, at times, misleading. Understanding that such companies actually draw on a long history of automatic music which did not start just decades ago, will help everybody gain a more balanced view of today's developments.

In the end, the story presented in this short and hopefully engaging volume is deeply personal in the sense that it filters the otherwise overwhelming amount of information through my own, ongoing research of the interplay of music and mathematics.

From continuities…

1
NOT A REVOLUTION (INTRODUCTION)

Creators and promoters of technology tend to label almost everything a "revolution." A camera's picture has become 15 per cent larger? A revolution.

Yet, technical change does not work this way. New, experimental technologies take extended periods of time for proposal, testing, building, and refinement, thus blurring, and often deleting, any clear temporal boundary that is supposed to be the "revolution." Moreover, the newest model of something just entering the market is often already obsolete from the internal point of view of the producer, as the next model is mostly already in preparation.[1]

Likewise, the use of AI for music is not a revolution, but the outcome of a long-term development. Both sides, computer technology and musical theory and practice, have formed over considerable time with much thought and work by innumerable actors. And while digital computers have started decades ago, music technology can look back at a history of millennia.[2]

Even some of the very first surviving artifacts of human history that we interpret as musical instruments (with the cautionary caveat that they, in principle, can also be something completely different) can be seen as an attempt to automate certain aspects of music-making through technology.

Extremely early prehistoric flute-like instruments made of bone have been found in recent decades. Some of them are at least 35,000 years old.[3] This is tens of thousands of years prior to any surviving or assumed traces of agriculture in the modern sense of the word, let alone writing. Yet,

DOI: 10.4324/9781003229254-3

already at this time, these bone flutes offered a technical means to replace certain mental and physiological functions of the human body that are needed for what we today call music.

What is the essence of a flute? Basically, a flute is a tube with holes. Blowing into the tube creates a tone, and opening and closing the holes changes the pitch of the tone.[4] However simple this might seem at first sight, the creation of an idea of a flute involves externalizing many processes that usually take place inside the body and mind, like the generation of audible vibrations of the air with the help of the vocal tract.

The bone flute helps with the physical creation of vibrations, replacing the glottis (the opening between our vocal folds) as the source of sound. But, crucially, it also replaces *mental* work. The player does not need to remember, and try to replicate, the pitch. The position of the holes has automated away this aspect of human activity.

From this moment on, more and more aspects of music will become formalized and automated over time. During a period of tens of thousands of years, musical instruments will be developed that will mechanize certain aspects of musical performance and also offer sounds that go beyond the possibilities of the human vocal tract and bodily sound-producing activities like clapping.

The pivotal situation when the creative activity of coming up with a musical piece (and not merely making it audible, as in the case of the musical instruments) starts to get partially mechanized, occurs around 1650.

In this year, the Baroque theorist and collector of scientific curiosities Athanasius Kircher publishes a treatise called *Musurgia Universalis* in which he proposes, among many other things, the use of a mechanical device to help with the composition of music.[5] It is important here that Kircher's wooden box with moving slats is solely meant to help with the *mental* work of remembering snippets of music and combining them, not with the physical task of creating audible vibrations. Equally important is that a real mechanism with moving parts is proposed, and not merely a system that works inside the human mind.

The way Kircher's proposed mechanism works is that slats containing multiple short snippets of musical material are arranged at random, providing a (hopefully) unique sequence of snippets which is then the resulting piece.[6] Kircher's approach makes use of the mathematical idea that the free combination of only a few snippets already yields a very large number of possible melodies (later, this insight was given the popular name of the "combinatorial explosion").

However, this systematic way of creating music is not the only possibility. Using the chaotic changes in nature as a source of unpredictability in music is another.

Again, it was Kircher, the avid collector of musical curiosities, who popularized this approach. In the same treatise, he described the idea of what is now commonly called an aeolian harp—a set of strings stretched over a frame and exposed to the natural unpredictability of wind gusts to produce an ever-changing, eerie sound landscape.[7]

In the aeolian harp, the combinatorial explosion mentioned above in relation to Kircher's box with slats is *not* the source of the richness of musical material created by the instrument. Even today, it is at least very difficult to understand the physics, acoustics, and perception of the sounds created by the apparatus in such a way as to describe it mathematically. Certainly, it was impossible during Kircher's time and in 18th and early 19th century when the aeolian harp had its moment of widespread popularity. Instead, the builders of aeolian harps relied on the changing wind to create variation in the device's sounds, mostly without even trying to really formalize the process. Therefore, such an approach to the automatic creation of music or sound is not mathematical in a strict sense.

A similar but much more controlled musical experiment is the wind chime—the set of tubes or sticks made to bump into each other by wind or by moving the chime. Normally, the tubes of the chime are not all the same so that their respective tones are also different, creating a bubbling acoustic play.

One could ask, why is the wind chime more "controlled" than the aeolian harp? Isn't the wind still unpredictable? It is, but the number of possible individual sounds that the apparatus can produce is restricted in the chime. More or less, each tube can only sound a certain tone, unlike the string of the aeolian harp whose acoustic subtleties seem to form an endless continuum.

By limiting the number and quality of possible tones, the builder of the wind chime thus takes care of a central principle of music-generating technology—the reduction of randomness.

It is easy to create a highly, almost completely, random sound. Just making a "shhh" noise is enough. It is also comparatively easy to create a sound that is so complex that the human listener is unable to hear the order in it, effectively making it seem completely random. Programs called "pseudorandom generators" are very good at it.

However, to create any kind of impression other than the sublime appeal of the Unintelligible, a certain *intention* needs to manifest itself in the sound that creates discernible musical *forms* that distinguish themselves from the chaos or uniformity of their *background*.[8]

This creation, and at the same time restriction, of unpredictability is a theme that will resound throughout the history of mathematical music.

Notes

1 Jonathan Sterne, "Out With the Trash. On the Future of New Media," in *Residual Media*, ed. Charles R. Acland (Minneapolis, London: University of Minnesota Press, 2007), 16–31.
2 Thor Magnusson, *Sonic Writing. Technologies of Material, Symbolic, and Signal Inscriptions* (London: Bloomsbury Academic, 2019).
3 Robert Anderson et al., "Archaeology of Instruments," Grove Music Online, 2020.
4 In the more precise sense, pitch is only the perceptual quality of frequency, which is the underlying physical phenomenon. However, this distinction will not be adhered to in this book for the sake of better readability and the avoidance of specialist terms.
5 Athanasius Kircher, *Musurgia Universalis Sive Ars Magna Consoni et Dissoni in X. Libros Digesta* (Rome: Corbelletti; Grignani, 1650), Vol. II, Book VIII.
6 John Zachary McKay, *Universal Music-Making: Athanasius Kircher and Musical Thought in the Seventeenth Century*. Doctoral Dissertation, Harvard University, 2013, 302.
7 Athanasius Kircher, *Musurgia Universalis*, Vol. II, Book IX. Marianne Bröcker, "Äolsharfe," MGG Online, 1994.
8 Abraham Moles, *Information Theory and Esthetic Perception* (Urbana: University of Illinois Press, 1966); Nikita Braguinski, *RANDOM. Die Archäologie der elektronischen Spielzeugklänge* (Bochum: Projekt Verlag, 2018), 163–181.

2
SINCE ANTIQUITY

The core methods and beliefs of mathematical music have been accumulating gradually since antiquity. Many ideas behind today's approaches have been around for differing amounts of time. Now, they are gradually coming together in a form that can be seen as something like a piece of writing that has been written over many times. Each of the many layers on this imagined document is visible, even if only barely. The specialist's name for such a source is *palimpsest*.

Deriving from with this image, I will treat the history of mathematical music as a kind of a palimpsest of ideas. It has formed as some historical approaches were overlaid with others—a process which almost never made the older ones disappear completely. Accordingly, the chapters in this part of the book will be called "Since Antiquity," "Since the Middle Ages," and so on to underscore the continuing relevance of centuries-old ideas even in today's mathematical music.

In antiquity, we find some of the earliest attempts to formalize mathematically the basic building blocks of music. The most famous example is the mystical-philosophical teachings of the mathematically inclined ancient Greek theorist Pythagoras who is reported to have lived during the 6th century BCE. His teachings, transmitted through later authors, contain a crucial musical–mathematical concept: ratio.[1]

Ratios are descriptions of relations. A ratio of 1:2 means that one thing is two times bigger, or wider, or louder, and so on, than another. Ratios are at the core of many cultural concepts. We find them in architecture,

DOI: 10.4324/9781003229254-4

sculpture, painting, even literature, and, of course, music and mathematics. Additionally, their use in culture is often connected to the idea that the human mind somehow derives special pleasure from being able to grasp the underlying mathematical order.

In principle, any relation expressed by two positive integers (whole numbers) is a ratio, in the everyday sense of the word. 1:2 is one, as is 38,146:69,531. However, in this example, only the former ratio can be understood and remembered easily and intuitively. Accordingly, only relations between relatively small numbers were historically regarded as the kind of a ratio likely to elicit a special aesthetic response.

The central idea that links ratios and music, a thought that has been permanently associated with Pythagoras since antiquity, is that such ratios made from small integers (especially 1, 2, 3, and 4) are the underlying hidden order enabling the aesthetic appeal of sound.

This aesthetic appeal is almost always linked to the idea of harmony. It has been often postulated that a harmonious acoustic impression is created in the listener when the sound is somehow structured by simple ratios. In the mystical-philosophical system of Pythagoreanism, and in later similar systems, these relations between small numbers are in fact expressions of a larger divine and cosmic order.[2] One example is the concept that proportions in music represent the proportions in the movements of celestial bodies—the so-called *music of the spheres.*

If we take a break at this moment to look back at the argument so far, we can see that it is deeply mathematical: hidden order manifests itself in form of properties that can be understood with the help of arithmetic. From now on, this idea will accompany our historical journey. The mathematical tools themselves will gradually become more complicated, but the core belief in the revelatory powers of mathematics to help with the creation of impressive aesthetic works will remain.

Also, it is worth stressing here that most mathematical musical theories are *speculative*. The authors of these theories believe, and are indeed often firmly convinced, that the internal logic of music works in some particular way which can be explained with the help of mathematics. Yet, *proving* a specific hypothesis by a specific author in such a way as to convince everyone is mostly impossible. In addition, speculative theories tend to be somewhat detached from musical practice and pedagogy of their time. Today, different traditions, schools of thought, and idiosyncratic approaches by individuals continue to coexist despite innumerous claims to "singular" or "universal" musical truth by various historical and contemporary theorists.

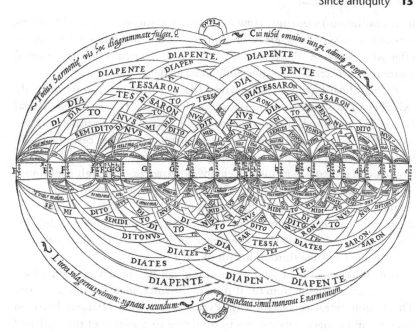

FIGURE 2.1   An illustration of ratios in a musical tuning system. From Francisco
de Salinas, *De musica libri septem*, p. 124. (Salmantica: Gastius,
1577).

In this situation, it is little wonder that music theorists, like their colleagues in the sciences, have often relied on the persuasive power of illustrations to mobilize support for their ideas. The simple ratios in the
historical theories of the Pythagorean tradition were a "natural" candidate
for this kind of visual argument. Such illustrations have sometimes grown
into very intricate depictions whose stunning visual beauty was obviously
meant to validate the musical speculation (see Figure 2.1).

The modern educated person is often accustomed to using certain
mathematical notions without thinking about their origin. Yet, consider-
ing an unfamiliar, historical perspective on them can help appreciate the
work that has led to the shaping of these ideas in the first place. For ratios
to become thinkable, the notions of a number, and of relations between
numbers, needed to exist.

A whole prehistory of gradual development of mathematical ways of
thinking has preceded Pythagoras and his theories. At some point in this
prehistory, the idea must have emerged that continuous qualities (such
as, for example, length) can be meaningfully represented by discrete

(non-continuous) elements. How long a road is, can be measured with increasing, almost endless, precision. But saying that its length is 100 steps is often *good enough*. At this point the road mutates from a continuous mental figure into a discrete one, sufficiently described by a number.

Ancient board games have survived that beautifully illustrate the early existence of this mental discretization of the world. The so-called "Royal Game of Ur," found during excavations in Iraq, is an exquisitely crafted wooden game whose principle is based on moving the pieces on a board which is divided into quadratic fields.[3] The player needs to move a piece along a certain path, but this movement is *discretized*: although continuous in the real world of the players who hold the pieces in their hands, it is thought of as consisting of "jumps" from one field to another. It is fascinating to consider that the modern division of musical pitch (tone height) into discrete steps is based on the same figure of thought as the Mesopotamian board game from the third millennium BCE!

Ratios based on such discrete thinking can be applied to different aspects of music ranging from the tuning of the instruments to musical form. The famous idea associated with Pythagoras is that relations between pitches can be made harmonious by adhering to a system of tuning based on proportions. In the canonical example, a string is shortened by half, by two thirds, and so forth, creating proportions of 1:2 or 1:3 between the pitches of the whole string and the shorter part. A second famous example associated with Pythagoras, which is, however, unlike the previous one, not really convincing from the point of view of physics, is that he was inspired by the sounds of different hammers.[4]

Building upon this approach, whole scales containing the notes from which the performer could choose to create a melody were calculated. Given the link to divine order described above, it is not surprising that, historically, such calculations were destined to be accompanied by claims of their superior aesthetic and mystical-philosophical qualities.

Mathematical attempts at rationalizing musical practice in form of scales and other calculation-based theories were not confined to the ancient Greek culture. In ancient China, for example, the Lü system of tuning brought mathematics into music in a similar way. Other impressive uses of mathematical devices in music can be found in various cultures of the world.[5]

Given such an early and global spread of the idea to use mathematics in music, one is tempted to see it as "natural." After all, it seems natural to base the tuning of the different strings of an instrument on ratios, especially now that modern physics has shown that vibrating strings simultaneously produce many *harmonics* (tones), and that these harmonics do represent ratios such as 1:2 and 1:3. And, certainly, ratios have played a central role in the centuries-long tradition of Western music theory, enabling a gradual build-up of a large and complex conceptual apparatus.

But does music theory *have* to rely on small-integer ratios?

From today's point of view there are several problems with this approach. First, it is now known that humans are unable to hear small deviations of pitch.[6] According to widespread assumptions, a ratio between the frequencies of two tones that is made up of small integers (like 2:3) will produce a harmonious result, whereas one made up of large integers will not. But a ratio such as 20,000,001:30,000,001 will, despite its monstrous numbers, yield frequencies that are so close to the ones produced by 2:3 that nobody will be able to distinguish between the two results.

Second, humans tend to perceive sounds (and everything else) using the categories they learned from previous examples. Therefore, a person who has repeatedly encountered a ratio like 2:3 in music, even without knowing about the musical theory behind the sound, is very likely to also perceive a slightly deviating sound as being the same—even if it is perfectly capable of hearing the difference when instructed to do so.

In addition, humans seem to have formed a common vocabulary of perception tactics such as grouping phenomena together according to similarity or continuation that are so strong that they even lead to manifold distortions of heard reality in form of auditive illusions.[7]

These aspects combined, the boundaries of perception and the tendency to "adjust" impressions to fit known models, put a question mark above the claims to "natural" legitimacy of musical theories relying on ratios. Yet, the fact that a theory is not "natural" does not prevent it from becoming widespread and influential. Over time, it can begin to feel natural to everyone who is accustomed to it.

Some ratio-centered theories have remained popular throughout the centuries since Pythagoras, and are used even today by those musicians and theoreticians who are especially interested in so-called "pure," or "just" intonation.

Pure intonation theories claim that only music based on tunings derived from small-integer ratios is truly harmonious and aesthetically satisfying. Their enemy is the tuning system commonly used today, which has abandoned simple ratios by making the distance between all tones equal.

The mathematical "problem" with the modern system is that the distance between tones is not just not expressible as a relation between *small* integers, it is not even expressible as a relation of integers at all (a so-called irrational number). Traditionally, scales of Western music were, and still are, theorized as further subdivisions of a big musical interval represented by the ratio 1:2, called the *octave*. A modern piano has twelve keys inside each octave, and, to make the distance between the tones equal, each tone has to be higher than the previous one by the twelfth root of two, which is an irrational number.

Simple ratios do still inhabit this modern system, but only as ideal models which today's tuning is capable of approximating, but not of reaching exactly. In the eyes of pure intonation musicians, the modern system is detuned. In the eyes of others, this is a neat observation, but one that is rather irrelevant in practical situations.

Furthermore, modern sound analysis technology shows that expressiveness in music is often linked to variations in pitch: an experienced singer or violinist, for example, is capable of making a tone a little too low or too high, or to play around with pitch in a myriad other ways, conveying subtle changes in emotion that would otherwise become lost. An artificially created "perfect" tuning of a singing voice with absolutely no variation in pitch (something that can only be achieved by technological means) is, by way of contrast, commonly perceived as mechanical.

Also, the idea that harmonious sounds are the building blocks of music leaves out the other important component: disharmony. It is probably impossible to create a musical impression of tension, drama, or of negative emotions, using only deeply harmonious and calming sounds. This observation is another problem for ratio-based ideas of music: even if a musician truly believes that only sounds based on small-integer ratios are pleasant, why not add in some unpleasant ones for the sake of dramatic development? In fact, descriptions of possibilities for tones to move from a more tense ("dissonant") to a more relaxed ("consonant") combination have been a staple of music theory for many centuries.

As can be seen from the description above, the issue of ratios in music is a multifaceted and difficult one. Still, it is obvious that the core idea that ratios are somehow centrally important has never disappeared. Some people are deeply invested in pure intonation, others simply remember that the octave is a proportion, still others play string instruments and therefore constantly have in front of them the equivalent of the Pythagorean string to be divided by 2, by 3 and so on.

In all these situations, the thought that music is deeply connected to mathematics is reinforced. Again, it seems completely "natural" to connect these two activities. And, again, we should remind ourselves that a different view of things is also possible, that music is, in fact, thinkable without mathematics.

Trying to understand why it has been consistently linked to the tools of mathematics is a topic of cultural history that exceeds the scope of this book. But I would like to at least offer one explanation that seems probable to me. My stance is that music was often used as a vehicle to show the effectiveness of some idea or method in areas *other* than music.

Arithmetic is a powerful tool. It can help measure land, construct buildings of needed volume, predict the movements of celestial bodies, or plan the agricultural season. By taking harmonious musical sounds as something (seemingly) obvious and natural, and by linking them to arithmetic operations such as multiplication or division, the Pythagoreans made a case for the usefulness of their mathematical knowledge in all areas, not just in music.

This demonstration of the power of mathematics through its application in music is another theme that will constantly accompany us on our journey through history. In 17th century, the Baroque theorist Kircher would use variations of his wooden box not only for composition, but also for encryption.[8] Later, 21st-century AI based on machine learning would be capable of segmentation and generation not only of musical scores, but of all kinds of information. Having a wooden box, or a machine learning algorithm, help with the composition of music means that these ideas and technologies do actually work, and can be applied in more "serious" situations.

Notes

1 André Barbera, "Pythagoras," *Grove Music Online*, 2001; Manuel Pedro Ferreira, "Proportions in Ancient and Medieval Music," in *Mathematics and Music. A Diderot Mathematical Forum*, ed. Gerard Assayag, Hans Georg Feichtinger, and Jose Francisco Rodrigues (Berlin: Springer, 2002), 1–26.
2 Catherine Nolan, "Music Theory and Mathematics," in *The Cambridge History of Western Music Theory*, ed. Thomas Christensen (Cambridge: Cambridge University Press, 2006), 273.
3 "The Royal Game of Ur," The British Museum, https://www.britishmuseum.org/collection/object/W_1928-1009-378; Irving L. Finkel, ed., *Ancient Board Games in Perspective. Papers from the 1990 British Museum Colloquium* (London: British Museum Press, 2007).
4 Calvin M. Bower, "The Transmission of Ancient Music Theory into the Middle Ages," in *The Cambridge History of Western Music Theory*, ed. Thomas Christensen (Cambridge: Cambridge University Press, 2006), 142–143.
5 Nick Collins, "Origins of Algorithmic Thinking in Music," in *The Oxford Handbook of Algorithmic Music*, ed. Alex McLean and Roger T. Dean (Oxford: Oxford University Press, 2018), 74–75.

6 Gareth Loy, *Musimathics: The Mathematical Foundations of Music* (Cambridge; London: MIT Press, 2006), Vol. 1, "6.4 Pitch", 156–166.

7 Diana Deutsch, *The Psychology of Music* (San Diego; London: Academic Press, 1999), "Grouping Mechanisms in Music", 299–348. For an example of an *optical* illusion, see the circular diagram of connections between letters in this book. The area inside the circle seems to be darker than the surrounding area.

8 Siegfried Zielinski, *Deep Time of the Media. Toward an Archaeology of Hearing and Seeing by Technical Means* (Cambridge, MA: The MIT Press, 2006), 141–157.

3
SINCE THE MIDDLE AGES

The Middle Ages bring about an intermediary step in the history of mathematical music. It first happens outside of music itself, but it will influence it several hundreds of years later. This crucial new development is the use of *combinatorics* for the purposes of communication (later also including aesthetics).

Combinatorics is a branch of mathematics concerned with counting and listing possible arrangements of individual elements.[1] For example, it can count all possible pairings of A, B, and C (there are nine if the order of the elements in the pair is essential), and it can list them: AA, AB, AC, BA, BB, BC, CA, CB, CC.

Combinatorics itself is older than its medieval use discussed here. Since antiquity, combinatorial approaches emerged in different cultures around the world. One of the most famous examples is the Chinese treatise *I Ching*, which is structured around a basically combinatorial idea: variations of a graphical symbol are created from six stacked lines, each of which can be either broken or unbroken. Given the two possible states for lines (broken and unbroken), and the number of lines (six), a total of sixty-four symbols is presented in the book and interpreted in relation to divination practices.[2]

Combinatorial approaches create large amounts of material. The sixty-four possibilities in *I Ching* are actually a rather modest example of the numbers that can be achieved with just a handful of initial elements. The medieval Christian mystic, author and theorist Ramon Llull (ca. 1232–1316)

DOI: 10.4324/9781003229254-5

appropriated this method to the generation of religious statements, effectively bringing combinatorics into the realm of communication.

His idea was to formalize the process of coming up with sentences. The first step was to let letters of the alphabet represent certain building blocks of language from which sentences are constructed. By doing so, he divided form (sentence structure) from contents (the respective religious concepts in his case), enabling further mathematization of argumentation. The next step was to postulate that letters could be chosen randomly, bringing into play the power of combinatorics. A graphical representation of his system in the form of circles inscribed with the letters was the crucial final step that both made the method easier to use and proved visually its power by the many lines connecting the letters.[3]

Figure 3.1 shows sixteen letters, arranged in a circle, with each letter connected to all the others. It is derived from an illustration in a medieval manuscript of one of Llull's systems.[4] The sheer number of connecting lines and the intricate symmetry of the figures that they form turn Llull's circular diagrams into visually impressive, enigmatic symbols reminiscent

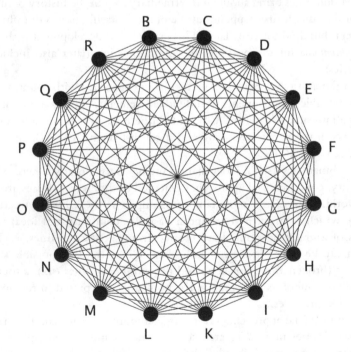

FIGURE 3.1 A graphical representation of all possible connections between sixteen letters.

of the *mandala*—a round figure used for meditation in Eastern religious traditions.

Yet, the diagram also works as a practical tool for the combinatorial creation of pairs of notions, to be used as material religious (in Llull's case) writing and arguments. In the original manuscript, each letter represents one divine attribute such as goodness, greatness, or eternity.[5] The circular diagram thus clearly suggests that each of them can be considered in relation to any other one, and that there are numerous possibilities for doing so.

By creating this combinatorial procedure for communication, Llull basically hoped to make interreligious arguments and conversion as easy as ABC (or BAC, or CAB...). Yet, his most important contribution was that he demonstrated the conceivability of mechanized thinking and communication between humans. From there, it was later only a small step to try to mechanize the process of coming up with melodies.

Llull's approach can be subdivided into several constitutive parts:

• The combinatorial idea that arrangements of a finite number of elements can yield a virtually endless richness of material.
• The idea to use a circular arrangement of elements.
• The idea to also use multiple circles to help with the mental work of combination.

Arranging letters in a single circle already enables geometric play that can be aesthetically interesting as well as symbolical in a mystical sense.[6] For example, one can choose letters in such a way that the lines that connect them represent a recognizable figure such as a triangle or a square. This figure, in turn, can be seen as representing a religious concept. The longer the sequence of letters, the more intricate will be the resulting path through the circle. At some point, tracing it can become a meditative, unworldly experience.

If one wishes to stay more alert during the mind-numbing work of combining strings of letters, one can employ a very simple, but powerful mechanism: multiple circles placed inside each other. With it, creating and remembering a sequence of letters is easy. Each circle, rotated into the needed position, will stay that way, helping with the construction of the arguments that Llull hoped to automate. Historical manuscripts of Llull's works contain such combinatorial aids, with multiple circles of different sizes on top of each other, all held together by a string going through the page.[7]

Like the holes of the prehistoric bone flute that mechanized the mental work of remembering pitch, these movable circles mechanize the work of remembering and combining letters and the concepts that they represent in Llull's system. And even if one does not cut out and assemble the circles into a thinking machine with movable parts, the visual representation of the system on a sheet of paper already makes the combinatorial work much easier than it would have been when conducted entirely in imagination.

Various historical sources have been suggested for Llull's method, including Jewish and Arabic influences. Notable parallels in its reliance on combinations of letters exist to the Jewish cosmological treatise *Sefer Yezira* which in the 12th century became a staple of the then-new Jewish mystical tradition of Kabbalism. *Sefer Yezira* enumerates the possible combinations of two letters of the alphabet, and it even contains the more general combinatorial calculation of the number of possible arrangements of up to seven letters. Likewise, Arabic *za'irajah* "letter magic" which is based on operations with circles inscribed with letters, and which has been later described in detail in the treatise *The Muqaddimah*, comes remarkably close to Llull's circular arrangements.[8]

Music seems predestined for combinatorial exploration. *Sefer Yezira* already lets us know that just seven different elements can already be rearranged in 5,040 ways. Accordingly, a musician is able to create this astounding number of melodies from just seven different notes. This is an inspiring and somehow liberating idea. It mitigates the fear (known to many creative individuals) of not being able to come up with *anything*. It feels great to know that fourteen notes already give rise to much more possible melodies than there are people in the world, without even taking the melody's rhythm into consideration![9]

Of course, several difficulties arise. Listeners might not accept most of these melodies either because they contradict written and unwritten rules of musical tradition or because they seem to represent no clear musical gesture that one can relate to. They can also grow tired of the uniformity of the musical system's output. And probably, they would see a melody written by a real person as more valuable, even if it happened to be the same sequence of notes as one of the millions of combinations produced by an automaton.

Still, combinatorial play is fascinating. The unusually large numbers of possibilities created by combinatorial methods are deeply counterintuitive. It is difficult to believe that shuffling around a small group of elements (which we can count and clearly imagine in our mind) gives rise to thousands, millions, and billions of combinations, counting which obviously exceeds our mental capabilities. This circumstance has been called *combinatorial explosion*.

Being an "explosion," it carries in itself not only positive connotations such as the unexpected richness of produced material, but also hints of danger: the amount of automatically created artefacts can be too much to manage effectively, or it can destabilize the cultural production economy by devaluing "manually" produced works.

By bringing a mechanical aid into the realm of human creativity (in this case, the creation of arguments), Llull therefore transgressed a boundary which is still likely to cause irritation. Missionaries and theologians of Llull's time were probably as unwilling to be automated away as are today's factory employees, office workers, or musicians. Even unpaid creative work conducted purely for recreation might start to feel less attractive to some people if they were to learn that a machine is capable of producing very similar results. Exaggerated versions of such feelings seem to have recently erupted in popular culture and everyday discourse in form of doomsday scenarios of machine domination and the ensuing humiliation and enslavement of humanity by a future form of AI.[10]

In the end, Llull's idea has probably failed in religious disputes, but it proved influential in broader culture. During the so-called early modern period which followed the Middle Ages, Kircher was deeply inspired by it, as was the famous mathematician Gottfried Leibniz and many others. The next step after Llull introduced combinatorics into communication was to actually employ it in the creation of cultural artefacts. The early modern period was to become the time when people began to see some of the promises of automatic composition based on mathematics.

Notes

1 To ensure greater readability, the word "arrangement" is used here not according to its strict mathematical definition as an ordered selection from a set of elements, but in a more everyday sense.
2 Andrea Bréard, "China," in *Combinatorics: Ancient and Modern*, ed. Robin Wilson and John J. Watkins (Oxford: Oxford University Press, 2013), 66.
3 Teun Koetsier, "The Art of Ramon Llull (1232–1350). From Theology to Mathematics," *Studies in Logic, Grammar and Rhetoric* 44, no. 57 (2016): 55–80; J. Gray, "Computational Imaginaries. Some Further Remarks on Leibniz, Llull, and Rethinking the History of Calculating Machines," in *Dia-Logos: Ramon Llull's Method of Thought and Artistic Practice*, ed. Amador Vega, Peter Weibel, and Siegfried Zielinski (Minneapolis: University of Minnesota Press, 2018), 293–300.
4 Reprinted in Donald E. Knuth, "Two Thousand Years of Combinatorics," in *Combinatorics: Ancient and Modern*, ed. Robin Wilson and John J.

Watkins (Oxford: Oxford University Press, 2013), 15. The original illustration additionally has the letter A in the center.

5 Donald E. Knuth, "Two Thousand Years of Combinatorics," 15.

6 Teun Koetsier, "The Art of Ramon Llull," 63.

7 Anthony Bonner, *The Art and Logic of Ramon Llull. A User's Guide* (Leiden; Boston: Brill, 2007), 60–61.

8 Teun Koetsier, "The Art of Ramon Llull," 76; J. Gray, "Computational Imaginaries," 297–298; Joseph Dan, *Kabbalah. A Very Short Introduction* (Oxford: Oxford University Press, 2006), 15–18; Victor J. Katz, "Jewish Combinatorics," in *Combinatorics: Ancient and Modern*, ed. Robin Wilson and John J. Watkins (Oxford: Oxford University Press, 2013), 110–113.

9 The factorial of 14 is 87178291200.

10 Mark Coeckelbergh, *AI Ethics* (Cambridge, MA: MIT Press, 2020), 11–17.

4
SINCE THE EARLY MODERN PERIOD

In 1636 the French monk and at the same time Renaissance-style polymath Marin Mersenne published his book *Harmonie universelle*. It was a truly encyclopedic collection of the foremost ideas of its time on music and sound, including acoustics and, notably, combinatorics.

Following in Llull's steps, and equipped with a more advanced mathematical theory, Mersenne strived to provide the musician with new, powerful tools. His fascination with the combinatorial explosion is visible on many pages of the *Harmonie universelle*. Researchers studying Mersenne's work are often astounded and bewildered by the sheer scope of his tables showing the results of combinatorial exercises. For example, Mersenne first lists all 720 permutations (rearrangements) of six notes, filling four pages of his book with a uniform stream of "ut, re, mi, fa, sol, la," "ut, re, mi, fa, la, sol" et cetera, only to fill the following twelve pages with the same material in musical notation.[1]

Unlike Llull, who lived three centuries earlier and was basically self-taught, the educated Mersenne employed surprisingly complex combinatorial devices not only to list, but also to count the many possibilities offered by mathematical tools. One of these overviews tells the reader how many different melodies can be constructed from nine notes if several notes are repetitions of a previously occurring note (e.g., there are 60,480 if three notes are the same, and 1,512 if one note occurs twice and another one occurs five times).[2] *Harmonie universelle* arranges the results of such calculations in visually impressive tables that generally start with smaller numbers

DOI: 10.4324/9781003229254-6

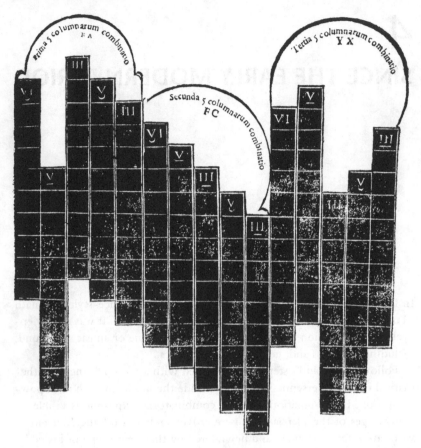

FIGURE 4.1   Combination of moving slats in Kircher's musical apparatus. From:
Athanasius Kircher, *Musurgia Universalis Sive Ars Magna Consoni et
Dissoni in X. Libros Digesta* (Rome: Corbelletti; Grignani, 1650),
Vol. II, p. 187.

at the top and gradually grow to unimaginably large numbers at the bottom, combining the persuasive power of a graphical illustration with the
mathematical evidence.

Mersenne's demonstration of musical mathematics did not remain a
solitary phenomenon. In 1650, Athanasius Kircher, another transitional
figure on the borders of religion and what is now called science and the humanities, published his *Musurgia universalis*. This two-volume work is even
more multifaceted than Mersenne's *Harmonie universelle*, covering all kinds
of musical and sound-related topics from the point of view of acoustics,
aesthetics, and, again, combinatorics.

It was Kircher's crucial contribution to the history of mathematical music that he proposed its *actual* mechanization, meaning the use of a physical device with moving parts for musical composition.

Figure 4.1, taken from Kircher's *Musurgia universalis*, shows the basic principle behind his musical apparatus. It illustrates three different random arrangements of five slats containing snippets of music. The arrangements are marked by arcs at the top. In the first arrangement, the second slat (inscribed with "V" in its uppermost square) has been shifted four squares down in relation to the first one, the third slat was moved one square up, and so on. Each arrangement offers several horizontal rows of squares which together yield the sequence of snippets. Previous chapters of *Musurgia universalis* provide the musical material to be written into the squares, and the readers are encouraged to build their own copies of Kircher's musical thinking machine.

Kircher's innovation was prepared by trends in both music and mathematics. On the musical side, there existed elaborate instructions on how to make *canons*, melodies that could be layered upon themselves (this means that a version of the same melody plays in addition to itself, but with a delay). Since their creation amounts to a kind of a mathematical problem, these instructions also became very formal and almost program-like. On the mathematical side, first mechanical calculation apparatuses emerged, such as an early form of the slide rule (a simple device that allows to carry out mathematical operations by moving its parts).[3]

Writing about his apparatus, Kircher makes two remarks that were destined to become a mainstay of authors' descriptions of their musical games and composition aids for centuries to come.

First, he posits that, by using the apparatus, everyone, even the most unmusical person ("amusos") should be able to create compositions that reflect the accepted rules of music theory without actually knowing them.[4] Implicitly, Kircher states here that those rules form a clear and coherent system that is so logical in itself that it can even be formalized as a physical tool. Despite the successes of modern music technology, the readers of today are well-advised to remind themselves that this is actually a contestable proposition, as there is still no universally accepted theory of music, or even of the relatively narrow stylistic range of music that Kircher could have had in mind in 1650.

Second, Kircher tries to impress the reader with the sheer magnitude of the numbers involved. He finds a very poetic description for the unimaginably large number of possibilities offered by combinatorics. He writes that even an angel who would have started progressing through the combinations at the moment the world was created, would not have exhausted them "to this day." Two centuries later, an advertisement for a Victorian composition machine would put it more bluntly: "IS IT TRUE? YES! Go and see for yourself. 428 Millions of Quadrilles, in one work," adding: "Price 5s. 6d."[5]

Kircher's tool for thinking musically can be seen as an extension of already existing technologies of mental work. Writing has been analyzed as an activity that has deeply restructured the way people think, even in situations of spoken communication.[6] Likewise, it can be presumed that the activities of writing and reading musical scores have restructured the theory and practice of music. At the very least, they have externalized the previously purely mental task of remembering the piece. And, probably, they have facilitated the creation of more complex, longer works. Even the notion of a work as a finished, authored piece whose form is fixed by notation (or, later, recording) probably could not have emerged without the cultural technique of musical writing.

At the same time, Kircher's apparatus complicates the scene by offering not one work, but the possibility of many works. In a sense, the machine itself starts to occupy the spot previously reserved for the cultural artefact. Unlike the author of a "traditional" piece who makes final decisions about notes or chords, Kircher merely sets boundaries around what the system can and cannot do. However, he, too, makes a lot of stylistic decisions, not only concerning the "hardware" of the box with its slats, but also the "data" of the musical snippets and the "software" of the operating instructions.

The practical test of Kircher's written descriptions from *Musurgia universalis* was finally accomplished when several copies of the apparatus were built in 17th century.[7] Like in today's programming, where only the actual running of the code is normally seen as a proof that it is error-free and indeed does what it is supposed to do, Kircher's machine had to be built and experimented with to show its value. Luckily, some of these apparatuses (which offer varying degrees of similarity with Kircher's descriptions) have survived. Copies of their contents have been made available in digital form, making it possible for everyone to assemble and "run" the historical machine and to form one's own opinion about it.[8]

Yet, even after "running" a copy of Kircher's machine (and overcoming its numerous ambiguities and inconsistencies that have been pointed out by researchers who did experiment with it) one is still one step removed from music as an auditive phenomenon. After all, the output of Kircher's system is mere notation, not sound.

Normally, this would mean in Kircher's time that the user would have to sing or play the resulting piece to finally turn it from information into vibration. Today's computer-based music systems easily combine the two tasks of generating a notated piece and playing it. But, interestingly, it was already possible to have a fully mechanized "production line" reaching from the creation of notation to playback using exclusively the tools of Kircher's era: In the same book, he describes the use of what he calls a "phonotactic cylinder," which is basically a pinned barrel activating a musical instrument mechanically instead of playing it manually.[9] Using this additional tool would, in principle, also remove the instrumentalist from the scene, in addition to the composer.

The next step in the development of composing apparatuses, after their principles have been laid out in the literature, was their popularization and commercial success. Two systems from the second half of the 18th century stand out in this respect: One, rather less known, because it was one of the earliest, and possibly the earliest such publication, and the other because it became so famous that it is still widely remembered as a precursor to all modern composing systems. The basic idea behind both systems is, however, the same.

The earlier publication is Johann Kirnberger's *Der allezeit fertige Polonoisen und Menuettencomponist* (The ever-ready minuet and polonaise composer) from 1757.[10] For each of the three musical genres the system is capable of imitating, it contains a "score" in which each bar is individually numbered. This "score" is not meant to be played from beginning to end. Instead, the sequence of bars is to be determined through random numbers yielded by throwing dice. Kirnberger is flexible with regard to the "hardware" needed to use his system and provides instructions for musicians with just one, as well as for those with two dice (if only modern computer programs were so undemanding!).

For the polonaise, Kirnberger offers 154 individual bars. Since a normal six-sided die only offers six different numbers from one to six, and even the combination of two dice (the sum of the dots) yields just eleven different possibilities, every number of dots is assigned several different bars in the "score." Kirnberger makes sure that the snippets follow each other in a more or less logical musical succession by prescribing that each bar of the resulting piece should be chosen only from a certain subset of the "score." In this regard, he goes beyond the basic idea of Kircher's apparatus where it was generally possible to form a string of self-contained snippets which were all interchangeable and could therefore stand at the beginning, in the middle, or at the end of the piece. In a sense, Kirnberger created a whole factory of Kircher-style systems: one for generating the first bar of the piece, one for the second bar, and so on.

Kirnberger's more complex system was made possible and manageable by a change in musical style that occurred at that time. Dance, with its predictably repeating musical form and formulaic melodic language, had become a primary example of a new musical aesthetics that foregrounded periodicity and simple pairings.[11] By making a dance (the polonaise) the basis of his system, Kirnberger inherited from it a rigid musical structure. This structure in turn facilitated the random choice among similarly written bars while, at the same time, offering a somewhat larger sense of long-term musical dramaturgy and development.

At the end of the introduction to his system Kirnberger puts a long quote written by an "ingenious artist of measurement," Mr. Gumpertz, explaining different kinds of combinatorial calculations and attempting to enumerate the possibilities offered by the *ever-ready composer*. It guides the reader through three mathematical formulas but finally fails to provide the correct number of possible pieces, seemingly because of a lack of a common language between musicians and mathematicians.[12] This problem is typical of its time. Whereas the polymaths Kircher and Mersenne previously united specialist knowledge from music and mathematics, later authors benefited from more advanced mathematical and musical theories, but had to pay the price of dividing the knowledge up among multiple persons, and of the difficulties of interdisciplinary communication.

After Kirnberger's pioneering effort came a whole wave of similar publications. One of them became especially well-known: the so-called *Musikalisches Würfelspiel* (musical dice game), historically attributed to Mozart. Its full title reads as *Instruction To compose without the least knowledge of Music so much German Walzer or Schleifer as one pleases, by throwing a certain Number with two Dice*. This system's lasting fame is not simply the result of minimal improvements over Kirnberger's approach.[13] The fascinating contrast between the genius image of Mozart and the mechanicity of the system must have contributed to its success. *Musikalisches Würfelspiel* gave everyone the possibility to become "Mozart" by creating a piece which is purportedly authorized by his name but still clearly made by the user him- or herself.

This interest in the mathematization of composition which is visible in the popularity of the "Mozart" system was not an isolated phenomenon. It was preceded by the rise of a whole musical aesthetics based on mathematical operations.

In 1739, the mathematician Leonhard Euler published a work that attempted to explain the perception of musical language through the listener's implicit mental computation.

In his *Tentamen novae theoriae musicae* ("An Attempt at a New Theory of Music") he proposed to grade musical *intervals* (combinations of tones as expressed by the distance between them, often written as a ratio) on a spectrum of more and less pleasant sounds. This was a significant deviation from contemporary explanations of the inner workings of Western music. Previously, musical theory mainly divided intervals into just two large categories: consonances (agreeable sounds) and dissonances (disagreeable ones). Whether a specific interval belonged into the "good" or the "bad" category, was, however, subject to historical change and a matter of debate among theorists. Assertions of musical primacy of "more perfect" ratios such as, for example, those where one number equals the other plus one (the so-called superparticular ratios like 3:2) played a role. But, basically, consonance was a matter of tradition and Pythagoras-like numerology.

Euler's hope, by contrast, was to give the question of consonance a scientific foundation. He went beyond the simplistic but practical dualism of consonance/dissonance by trying to calculate the exact *degree* of agreeableness (*gradus suavitatis*) of a given interval. Euler came up with a somewhat involved set of formulas that, ultimately, failed to influence musical theory or practice in any direct way. Yet, their underlying assumptions seem to have imprinted themselves into the fabric of later mathematical aesthetics.

Euler began by arguing that intervals are pleasing to the listener when they represent a ratio that can be understood. Correspondingly, he believed the lack of perceivable order to cause displeasure. So far, this was common sense. But then came his important argument: "some ratios are perceived more easily than others."[14] Euler seems to have imagined the perception of musical intervals to be oriented first and foremost toward the reduction of mental work needed to understand them. Thus, for him, less mental effort in deciphering their inner order must have resulted in more of this basic, simple pleasure (which could then be additionally complicated by the intellectual desire of complexity).

By proposing the use of his mathematical formulas to determine the agreeableness of intervals, Euler basically stated that the kind of mental work the mind does when perceiving sounds is *calculation*.

In this regard, he was supported by the writings of another theorist of his era, Gottfried Leibniz. The musically interested Leibniz has made several statements about the perception of tones which all boil down to the concept of unconscious counting. In one work, written before Euler's *Tentamen*, Leibniz, for example, made the argument that the beauty of music "consists [...] in a calculation, which we do not perceive but which the soul nevertheless carries out." This imagined mental mathematical dissection of sound shifted the perspective on intervals. Instead of being more *perfect* (which is their inner quality), consonant intervals were now thought of as more *intelligible* (a quality that only arises during their perception by a listener).[15]

Philosophy, too, offered treatments of aesthetics that strove to account for the perception of formal, "mathematical" designs. Important parts of the philosophical work *Critique of Judgement* (1790) by Immanuel Kant are concerned with understanding the "feeling of pure beauty" and its reliance on the degree of "potential knowability of a given object," again emphasizing the amount of mental work needed to classify a perception. Interestingly, Kant seems to have believed that the perception of abstract or purely formal designs worked in absolutely the same way in every person.[16] Ultimately, this idea means that the mathematical music's power to communicate should be even greater than that of non-mathematically created works.

At the same time, the emerging science of sound—acoustics—also seemed to support the idea of listening as counting inherent in Leibniz's and Euler's theories. Already at the end of the 17th century it has been proven by experiment that the pitch of a tone is a direct perceptional representation of its frequency. In other words, a higher number of vibrations per second caused the sensation of a higher note. Moreover, the then new experimental set-up consisting of a wheel with 'teeth' fixed to its outer rim to produce audible clicks when rotated made it possible to state *exactly* how many vibrations correspond to a certain perception.[17] The theorists of the 18th century were confronted with the fact that the mind somehow "knew" that the sound with more vibrations per second is higher, even without being consciously aware of the number of individual vibrations. The idea of an unconsciously counting mind (or "soul") clearly suggested itself in such a situation.

Overall, it gave rise to a new credibility of mathematical methods in music: If the listener is pleased by the inner order of sounds, it seems obvious to try to create this order through formal procedures.

During the following 19th century mathematical, physical, and physiological perspectives on sound continued to become more refined, while at the same time undergoing (again) a somewhat unexpected process of fusion with a magical mode of thinking, in effect preparing the ground for the triumph of today's mathematical and computer-aided methods in music.

Notes

1 Catherine Nolan, "Music Theory and Mathematics," 284; Eberhard Knobloch, "The Sounding Algebra. Relations Between Combinatorics and Music from Mersenne to Euler," in *Mathematics and Music. A Diderot Mathematical Forum*, ed.

Gerard Assayag, Hans Georg Feichtinger, and Jose Francisco Rodrigues (Berlin: Springer, 2002), 29–31; Marin Mersenne, *Harmonie universelle, contenant la theorie et la pratique de la musique* (Paris: Sebastien Cramoisy, 1636), livre second, 111–128.

2 Eberhard Knobloch, "The Sounding Algebra," 31; Marin Mersenne, *Harmonie universelle*, livre second, 130.

3 Sebastian Klotz, *Kombinatorik und die Verbindungskünste der Zeichen in der Musik zwischen 1630 und 1780* (Berlin: Akademie, 2006), 15, 16, 35.

4 Athanasius Kircher, *Musurgia Universalis*, Vol. II, 185. German translation: https://www.hmt-leipzig.de/home/fachrichtungen/institut-fuer-musikwissenschaft/forschung/musurgia-universalis/volltextseite

5 Athanasius Kircher, *Musurgia Universalis*, Vol. II, 188; Nikita Braguinski, "'428 Millions of Quadrilles for 5s. 6d.': John Clinton's Combinatorial Music Machine," *19th-Century Music* 23, no. 2, Fall 2019: 86–98, https://doi.org/10.1525/ncm.2019.43.2.86

6 Walter J. Ong, *Orality and Literacy. The Technologizing of the Word. With Additional Chapters by John Hartley* (London: Routledge, 2012), 77.

7 John Zachary McKay, *Universal Music-Making*, 302–305.

8 Wolfenbütteler Digitale Bibliothek, http://diglib.hab.de/?objekte=90-aug-8f

9 Athanasius Kircher, *Musurgia Universalis*, Vol. II, 312; Thomas Patteson, *Instruments for New Music: Sound, Technology, and Modernism* (Oakland: University of California Press, 2016), 24.

10 Johann Philipp Kirnberger, *Der allezeit fertige Polonoisen- und Menuettenkomponist* (Berlin, 1757), http://mdz-nbn-resolving.de/urn:nbn:de:bvb:12-bsb10527349-7

11 Wolfram Steinbeck, "Würfelmusik," MGG Online, 1998, https://www.mgg-online.com/mgg/stable/12552

12 The third formula given by Kirnberger in his quotation, "m to the power of n," is the one to be used in the case of his system. Polonaises containing a maximum of 14 different bars can be created with the help of the provided "score." The correct number of possibilities for such fourteen-bar polonaises using two dice is eleven to the power of fourteen because there are eleven possibilities to choose from for each bar. There are even more possibilities if one takes into consideration the common practice of repeating whole parts of such pieces.

13 A notable change compared to Kirnberger's system is the reduction of the number of voices from four to just two, making the piece more easily playable on a keyboard instrument and reducing the amount of preparatory copying work.

14 Peter Pesic, *Music and the Making of Modern Science* (Cambridge, MA: The MIT Press, 2014), 135–137.

15 Peter Pesic, *Music and the Making of Modern Science*, 136 and 144.

16 Robert Wicks, "The Beauty of Universal Agreement," in *European Aesthetics. A Critical Introduction from Kant to Derrida* (London: Oneworld, 2013).

17 Sigalia Dostrovsky and John T. Cannon, "Entstehung der musikalischen Akustik," in *Hören, Messen und Rechnen in der frühen Neuzeit*, ed. Frieder Zaminer (Darmstadt: Wissenschaftliche Buchgesellschaft, 1987), 32–33.

5
SINCE THE 19TH CENTURY

What is the next logical step after the use of musical combinatorics has been established, and its mental application made easier by Kircher's box with moving slats? In 1821, Dietrich Winkel, an inventor of music technology, mechanized the very process of making random choices. This enabled him to create his spectacular Componium machine which united the randomization and the playback, conjuring up something like an early illusion of non-human creativity.

To the astonished visitor of the Componium exhibition in Paris, the machine presented itself as a semi-magical automaton capable of producing "innumerable" variations of a musical theme. Its imitation of human musical activity worked so well that the Componium was even suspected of containing somewhere within itself a hidden musician, like the infamous Mechanical Turk chess automaton, which was eventually exposed as a fraud.[1] However, the Componium's mechanical randomization of music was genuine.

Winkel, who was also the original inventor of the metronome,[2] came up with the ingenious idea to combine a music-playing mechanism with a sort of a lucky wheel whose unpredictable rotation provided the needed source of randomness. The playback mechanism was a fairly traditional refinement of the pinned barrel idea already described in print by Kircher almost two centuries earlier. The only significant change was that there were two barrels taking turns to play short snippets, and each containing

DOI: 10.4324/9781003229254-7

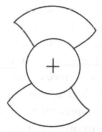

FIGURE 5.1 The Componium's randomization wheel. Based on Tiggelen, *Componium: The Mechanical Musical Improvisor*, Figure. 40.

several different snippets. But the Componium's mechanical randomness generator was its truly unique feature.

It worked by setting in motion a wheel in which two quarters of the outer rim have been removed, creating a butterfly-like shape (see Figure 5.1). Random choice depended on whether the wheel stopped on one of its filled or, rather, empty sides. Each possibility could thus occur with a 50 per cent chance. Depending on that, each barrel either repeated the same snippet when it was again its turn to play, or switched to the next one.[3]

Given that each snippet was very short—just two bars of music—and that the snippets from the two barrels alternated and repeated in an unpredictable manner, it is easy to believe that Winkel's contemporaries were very impressed by what seemed like a musician's free improvisation on a theme.

When thinking about such a situation, two difficult questions arise. The first one is whether a musician's improvisation is really fundamentally different from what the Componium accomplishes. On the one hand, human improvisation is driven, at least in part, by the fleeting momentary states of mind, by an individual artistic goal, and by the response of the audience and fellow musicians. This territory is clearly not covered by the Componium's procedure of randomized playback of preexisting snippets. Yet, improvisation, especially the one imitated by the Componium, is not completely free and unpredictable. Rules of musical tradition and genre constrict the range of possibilities. In certain genres like the historical dances, but also in 20th-century rock and jazz, popular formulas of melody, rhythm, and chord progression exist that are generally expected by the listener, and that additionally limit the range of possibilities. In the end, the "output" of an improvising musician might resemble, to a degree, the output of a well-constructed randomizing system based on the choice of snippets, even if the process that leads to it is completely different.

The second question is whether the listeners would still be impressed if they possessed superhuman memory and thinking capabilities which would make the Componium's whole mechanism so evidently clear to them that it would become uninteresting. A well-known enemy of musical pleasure (again, we are limiting our observations here to historical styles and social situations) is monotony. So, at which point would monotony have occurred if we were to simplify the scheme by which the Componium operates? Clearly, it would be the case if the same two bars were repeated endlessly with no change. But what if the Componium had just four different snippets, instead of dozens of them? It might be just enough to maintain the interest of some listeners who are new to the genre, comparable to the situation where they would be hearing the results of the efforts of an unskilled improvisor. A music professional would see the underlying mechanical structure more quickly, as a result of the training received in hearing, memorizing, and analyzing music. Eventually, the professional might glean the system's inner logic, and become bored, after a shorter amount of time. Now, switching back to the full Componium, and imagining a kind of a superhuman musicologist whose mental capabilities are equally larger, it is easy to see the ways in which combinatorial music depends on the limitations of our thinking, memory, and knowledge.

The Componium had one crucial advantage when compared to Kirnberger's system consisting of a list of bars and a set of dice: it worked in real time. Kirnberger's approach involved throwing dice, looking up a number in a table, copying the bar onto a separate sheet of paper, and only then playing it. It took time to construct a piece using Kirnberger's book, even if the reader were to cut out individual bars (as recommended in the system's description) and arrange them physically as a set of cards. But, on the other hand, the clear disadvantage of the Componium was its immense cost and mechanical complexity which has eventually led to it becoming unplayable after some time. A compromise solution was finally found around the middle of the 19th century by the British musician John Clinton.

Clinton's system, called the *Quadrille Melodist*, was relatively cheap in production as it only involved printing a certain amount of bars and cutting them into individual cards. Yet, it was also usable in real time, like the Componium. The way Clinton achieved this was by making the human pianist a part of the system, and by providing a practical sorting solution for the snippets that turned them into an immediately playable score.[4]

The copying part was removed by Clinton, as was the drudgery of endless dice-throwing and the need to consult a table. Instead, he came up with a simple tray with rows of pockets into which all the cards were sorted, depending on their specific place in the piece: The first pocket only contained cards for the first bar, and so on. The user was free to play the score formed by the cards at the top of the pockets as it was, or to switch cards whenever more musical variety was needed. The procedure of bringing random choice into the music was so easy (just pulling a card from a stack in the pocket and bringing it to the front) that it could be done between repetitions.

Figure 5.2 shows the general makeup of the *Quadrille Melodist's* sorting tray. In pocket 1, cards A1, A2, and A3 are stored, all representing possible choices for this part of the piece. Correspondingly, the next part is represented by the snippets of music contained on the cards B1, B2, and B3. Figure 5.2 is, however, only a simplified minimal illustration. The tray of the real *Quadrille Melodist* holds eleven cards in each pocket (A1...A11, B1...B11, and so on for its first piece, and an additional set of cards for its second piece) and has twenty-one such pockets.

Quadrille Melodist allowed for varying degrees of randomization. The users could theoretically decide to replace just one card before repeating the piece, making it sound as a very slight variation, or they could, instead, switch many or even all of the cards. As the cards were all numbered, it was even possible to always take the next card in all the pockets, effectively destroying the randomness (as card 1 of bar 1 would always be combined with card 1 of bar 2, and so on), and turning the system into a set of static variations, although this would obviously contradict its purpose.

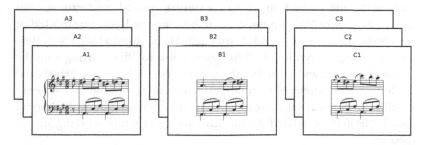

FIGURE 5.2 The structure of the *Quadrille Melodist.*

One difficult aspect of a combinatorial musical system like the *Quadrille Melodist* is that the creator has to take care of melodic continuity across cards. For example, the melody on the card in bar 1 should ideally be met by a seamlessly connecting melody on the card in bar 2, without unexpected jumping, abrupt changes in direction, or other inconsistencies that would contradict the historical genre of the piece. At the same time, just using the same melody on all cards of a certain bar with minimal variation would inevitably cause monotony after a couple of repetitions. Clinton therefore needed to construct the different musical snippets contained on the cards so carefully that they would more or less all fit all the preceding and following melodies. In doing so, Clinton might have taken a clue from another combinatorial pastime of his era, the Myriorama.[5] It was a set of cards with snippets of painted views of an imaginary landscape, which could be combined randomly while still creating a credible picture where slopes of mountains, for example, would continue logically across cards, like the melodies in Clinton's case.

This construction of a musical system where everything fits everything was an immense task, given that Clinton decided to use eleven cards in each of the twenty-one pockets. This decision, however, might have been the result of the same separation of disciplines that was already visible in Kirnberger's case (Kirnberger had to enlist the help of a mathematician to understand the combinatorics of his own system). Clinton, too, gives a number supposedly describing the scope of his system, and it, too, turns out to be incorrect. In fact, Clinton could have saved himself a lot of work by relying more thoroughly on the laws of combinatorics. With only three cards per stack instead of eleven he could already achieve a larger number of possible pieces than the (incorrect) amount that was used to advertise the *Quadrille Melodist.*[6]

Several decades after Clinton's *Melodist* is published in Britain, and already at the end of the 19th century, a German retired conductor sets out to save the fate of Western music by reinstating the Pythagorean ideal of pure proportion. His name is Eduard Grell. He has lived a long life in Berlin, witnessing everything the 19th century had to offer, starting with the Napoleonic wars, and including the political rise of his native Prussia and the formation of Imperial Germany. He also conducted the *Singakademie*, a well-known historical choir in whose famous building (now the Maxim Gorki Theater) he resided during his later years while working on his theories.

Grell's ideal was a mathematically "pure" system of tuning that was hoped to give back to the singer what Grell regarded as the only "natural" way of choosing the pitch of a tone. To this end, Grell devised several approaches, which all belonged to the realm of "just" intonation.[7] This tradition of relying on simple ratios for tuning has existed in some form since antiquity despite its numerous problems such as the inability of listeners to hear differences of pitch beyond a certain limit (see the chapter on antiquity in this book). Grell's grand goal was, too, to construct a new musical-mathematical framework which defines the steps of a musical scale in such a way that their frequencies are be related by a small-integer ratio.

Drawing on this "just intonation" idea that small-integer ratios have special aesthetic qualities, in the 1880s Grell created his various new tuning systems with up to sixty-four steps per octave, far more than the standard number of just twelve notes. It should be stressed, however, that he did so not in some rebellious attempt to overthrow traditional Western harmony and replace it with unheard-of microtonal melodies. Quite the contrary: Grell was a traditionalist, and his goal was to facilitate the use of just intonation in performances of works influenced but by what he considered the ideal of all "true" music—the solemn religious choral works of the 16th century.

After Grell's death, some of his writings, though not his tuning systems, were published by his pupil and main collaborator Heinrich Bellermann.[8] In the wake of modernism and fin-de-siècle aesthetics, they, however, failed to gain any traction and were largely ignored during the subsequent decades. Yet, the handwritten pages containing unordered sketches, calculations, and tables of tones for Grell's just intonation project were preserved through the decades following his falling into oblivion. They are now a living monument to the longevity of musical-mathematical ideas going back millennia to Pythagoreanism and antiquity.

Even more fascinating with regard to musical practice, and its relation to idealized theoretical systems, is that Grell's archive also contains a large set of tuning forks representing the frequencies of the tones in one of his systems.

Grell tried to create a set of tuning forks showcasing his division of the octave into sixty-four tones. He hoped to use them to educate singers, through the technical means of the sounding fork, to sing according to the "natural" tradition of just intonation. However, he failed to realize that there are limitations both to the precision of manufacturing the musical tools of his time, and, ironically, to the precision of hearing the musical intervals. Grell's "just-intonation" tuning forks are in fact significantly detuned compared to the ideal frequencies of his system.[9] Did Grell himself hear that his tuning forks were in reality more "impure" than the so-called equal temperament system to which he was so ardently opposed because it was not based on simple ratios? We may never know—but his attempts provide us with material that clearly illuminates the tensions between idealist approaches in music theory and the materiality of all musical practice.

While Grell was a professional musician, but an amateur in the field of mathematics or science, the opposite was the case for the famous Berlin physicist Hermann von Helmholtz. He was also interested in just intonation. But, more importantly, he wanted to attack long-standing questions of music theory like the understanding of consonance (see the chapter on the early modern period) with the whole power of then-modern mathematics, acoustics, physiology, and professionalized research institutions.[10]

The following overview of Helmholtz's argument about consonance of musical intervals illustrates the degree of refinement in musical-mathematical theories that already existed in the 19th century:

1   Sounds of musical instruments that have a recognizable pitch mostly consist of many simultaneous vibrations (partials) at different frequencies. Mostly, these frequencies are multiples of the one low frequency that is heard as the musical tone, and are not consciously perceived. (This was already known by Helmholtz's time.)

2   When two vibrations occur at similar frequencies, they start to reinforce and annihilate each other in an alternating pattern, forming an audible rhythm called *beating*.

3   Following in the tradition of Euler's *gradus suavitatis* (the degree of consonance), Helmholtz attempted to specify what he called the *roughness* of an interval. His mathematics and physics were, however, more advanced. Helmholtz calculated the roughness from the amount of beating occurring between partials of tones, while also postulating and taking into account mathematically that beatings at a certain frequency are perceived as especially displeasing.

4   The result of Helmholtz's calculations were numbers. To make them more intuitively understandable, he created a figure illustrating how the roughness is at its lowest between the tones that form well-known "pure" intervals of Western music, while it is at its highest when these intervals are detuned.[11]

Figure 5.3 shows Helmholtz's illustration of how the roughness of a musical interval (the height of the lines) changes with the distance between the two notes. Pairs of numbers indicate here the partials (individual vibrations) contributing to the beating which, in Helmholtz's view, ultimately causes displeasure.

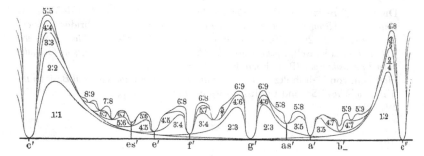

FIGURE 5.3 Helmholtz's illustration of the "roughness" of different musical intervals. (Helmholtz, *Die Lehre von den Tonempfindungen*, 1913, p. 318, figure 60a.)

Of course, Helmholtz's 19th-century theories did not remain uncontested. Still, the high standards of physical and mathematical thinking evident in his work on musical theory made it a very obvious idea that, to make progress in these areas, specialized research institutions were needed. Their eventual emergence was the defining feature of the next period, the 20th century.

Notes

1 Terrance Riley, "Composing for the Machine," *European Romantic Review* 20, no. 3 (2009): 375–376.
2 Terrance Riley, "Composing for the Machine" 374.
3 Philippe John van Tiggelen, *Componium: The Mechanical Musical Improvisor* (Louvain-la-Neuve: Institut supérieur d'archéologie et d'histoire de l'art, 1987), 315–316.
4 Nikita Braguinski, "'428 Millions of Quadrilles for 5s. 6d.': John Clinton's Combinatorial Music Machine," *19th-Century Music* 23, no. 2, Fall 2019: 86–98, https://doi.org/10.1525/ncm.2019.43.2.86. A copy of the Quadrille Melodist is held at the British Library (British Library Board, Music Collections M.1.)
5 Ralph Hyde, "Myrioramas, Endless Landscapes. The Story of a Craze," *Print Quarterly* 21, no. 4 (2004): 403–421.
6 Nikita Braguinski, "'428 Millions of Quadrilles for 5s. 6d.," 97–98.
7 Nikita Braguinski, "Die Systeme der reinen Stimmung von August Eduard Grell und ihr geistesgeschichtlicher Kontext," in *Jahrbuch 2011 des Staatlichen Instituts für Musikforschung Preußischer Kulturbesitz* (Mainz: Schott, 2011), 75–104.
8 Eduard Grell, *Aufsätze und Gutachten über Musik*, ed. Heinrich Bellermann (Berlin: Springer, 1887).
9 Nikita Braguinski, "Die Systeme der reinen Stimmung von August Eduard Grell," 92–100.
10 Sigalia Dostrovsky et al., "Physics of Music," *Grove Music Online*, 2001; Julia Kursell, "Hermann von Helmholtz und Carl Stumpf über Konsonanz und

Dissonanz," *Berichte zur Wissenschaftsgeschichte* 31, no. 2 (2008): 130–143; Benjamin Steege, *Helmholtz and the Modern Listener* (Cambridge: Cambridge University Press, 2012), 206–210; Peter Pesic, *Music and the Making of Modern Science*, 217–229; Julia Kursell, *Epistemologie des Hörens: Helmholtz' physiologische Grundlegung der Musiktheorie* (Paderborn: Fink, 2018).

11 Hermann von Helmholtz, *On the Sensations of Tone as a Physiological Basis for the Theory of Music*. Second English Edition. (London: Longmans, Green, 1885), 192–193.

6
SINCE 1900

A lot of people think that the roots of today's digital methods of making music go back just a couple of decades, maybe to the time around the middle of the 20th century—at most! We have already seen that this cannot be true, as the mathematical and formal-procedural approach to music that is the foundation of all such programming work has grown gradually over a much longer period. Now, let's take a look at the peculiar era at the beginning of the 20th century when almost everything needed for today's computerized approach was already in place, with the only thing still missing being the computer itself.

Helmholtz has left a lasting impression on theoreticians of music. Equally powerful was the impact of the renewal of musical language that occurred around the turn of the century. The combination of these two aspects formed the basis for a new development where avant-gardist overturning of tradition fused with a belief in science, mathematics, and technology.

Already in 1907 the pianist and composer Ferruccio Busoni published his *Sketch of a New Esthetic of Music*. In it, he argued for more experiments with non-traditional notation systems, microtones (intervals that are smaller than the ones commonly used in Western music) and the electronic synthesis of sound, combining new technology with new aesthetics, and explicitly referring to Helmholtz. In 1916, the avant-garde artist Luigi Russolo published his manifesto *The Art of Noises* which suggested a new aesthetics that confidently included industrial sounds, thus attempting to

DOI: 10.4324/9781003229254-8

remove the centuries-old boundary in Western music between "musical" and "unmusical" acoustic material.[1]

Probably the most well-known reformer of music in the first half of the 20th century was the composer Arnold Schoenberg. His claim to fame was a system of creating music using all twelve Western notes in the octave. Traditionally, only a subset of these twelve notes was used in any given piece due to the structure of familiar scales, chords, and melodies. Schoenberg proposed a method of avant-garde composition where the composer was led to abandon traditional musical language by a mathematical procedure that ensured all twelve notes were used equally.

This was a shocking musical experiment. It showed that a mathematical method can be used in the field of modernist aesthetics to create serious works of art, and not just for amusement, as was the case with the musical games of the 18th and 19th century.

Experiments by Busoni, Russolo, Schoenberg and others became forerunners of an international wave of musical change. Yet, as individually created visions they did not and could not compete with the huge collective work that was later carried out over the course of a decade in specialized research institutions. And it was due to historical and political reasons that the once relatively minor musical country—Russia—was to become the place where the new way of thinking erupted with particular power and consequence.

In 1917, a coup d'état carried out in Russia created a new entity that later became known as the Soviet Union, a totalitarian state with strict political control. Some of its early years, however, were marked by a temporary loosening of ideological constraints on life and thought, installed as a measure to fight off an otherwise inevitable collapse of economy and government. It was during these difficult and contradictory times when the early Soviet musical–mathematical avant-garde had its brief moment of blossoming, creating a tradition that continued to reappear throughout the West in the 20th century.

Between 1921 and 1928, a group of new Soviet research and teaching institutions deeply interested in a formal and scientific understanding of arts were formed, run, and also eventually closed on the regime's request. The use of science for art and music was their explicit goal. The very names of the institutes and schools made it clear that old sentiments not based on rationality were not welcome: *State Institute for Musical Science* (GIMN), *State Academy of Artistic Sciences* (GAKhN), or *Higher Studios for Art and Technology* (VKhUTEMAS).

During this time, the search for genuinely mathematical methods in music (as opposed to, say, intuitive methods based solely on inspiration) thrived in Soviet musicology. Building upon older approaches and with sustained, even if still meager, support by the new state, a group of mathematically inclined theorists set out to make Soviet musicology more substantial, and Soviet music better than anything that existed previously. All this ended abruptly when a new cultural policy centered on traditionalism was established by the Soviet state at the end of the 1920s, banning modernist theories for decades, and in some cases sending the researchers to deadly prison camps.

Still, a host of new theories managed to grow during these tumultuous times. Among these, two were of special importance to the use of mathematics in music.

The first one is the theory of proportions in musical form, proposed by Georgij Konjus (also written as Conus). Konjus extended the existing idea that ratios play a role in the perception of *intervals* into an attempt at understanding *large-scale* structures in music, such as whole movements of symphonies. He postulated that all masterpieces of music history were unconsciously fitted by their creators to some sort of symmetry which he, finally, was able to analyze explicitly through his novel methods.

Similar ideas were expressed by other early Soviet musicologists. For example, Leonid Sabaneev was convinced that all true masterpieces were based on the Golden Ratio, the so-called "divine proportion," and tried to prove this statistically by analyzing almost 2,000 historical works. Later, after escaping the Soviet Union, Sabaneev even went as far as to postulate in a mystical and fatalistic text that the Golden Ratio determines not only the formal structure of "good" music, but also the period of greatest achievement, as well as the moment of death of any person.[2]

Konjus chose for his theory the composite name of "metrotektonizm," alluding through the Greek words from which it is assembled to the widespread belief that architecture, too, is grounded in ratios and proportions, as well as to the idea that all arts are united by their underlying mathematics. Despite his nearly 1,000 analyses of scores, Konjus often had a difficult time convincing his fellow musicologists that the clean-cut symmetries he saw in the great works of music were in fact reflected in music itself, and the absence of attempts on his side to connect to the ruling ideology made him even less acceptable in official Soviet musicology.[3]

Still, Konjus and Sabaneev are historically important for taking seriously (maybe even a little too seriously) the promises of Pythagoreanism and Leibniz's "calculating soul" to explain mathematically the aesthetic qualities of music, and for trying to broaden the scope of mathematics from basic intervals to whole works.

The second notable development in Soviet arts research of the 1920s is the use of statistics. Statistics is a branch of mathematics concerned with collecting data on some large body of objects and presenting it in such a way as to facilitate the understanding of certain qualities or patterns that would otherwise remain hidden.

Already in 1913 the Russian (at that point he was not yet Soviet) mathematician Andrei Markov carried out an experiment in connecting statistics to arts. In it, he analyzed the sequence of vowels and consonants in a famous text of Russian literature.[4] The result of this endeavor was the popularization of the mathematical method now known as the *Markov chain*. It has since become a staple of computer-based music and arts experiments, and its appeal has not waned during the ensuing more than hundred years. Afterwards, simpler but still powerful statistical approaches to measuring historical change of style were experimented with by literary scholars, as well as by Sabaneev who tried to track numerically the frequency of chords and other elements of music to achieve this goal.[5]

The basic idea of the Markov chain method as it is understood today is to calculate how often a certain element (e.g., the chord G) is followed by other elements (say, chords C and C#) in a given source, expressed as probabilities. Having created this set of data, one can then use the information to generate new sequences of elements that would be quite similar, but not identical to the original. Of course, the cumbersome task of collecting data and doing calculations is greatly facilitated by the computer, as opposed to Markov's heroic attempt that was carried out by hand. Today, the Markov chain is—in a sense—the perfect method of musical mathematics as not only the application of rules to create a piece, but also the establishment of these rules in form of probabilities turns out to be a completely formal, unambiguous, and transparent process.

When modernist aesthetics and speculative mathematical inquiry into arts were banned by new Stalinist cultural politics in Russia, some individuals interested in these matters were lucky to be able to emigrate. Among them, one of the most successful was the composer and theorist Joseph Schillinger.

Originally a well-known figure of Soviet avant-garde, after his emigration Schillinger taught mathematical methods to his American pupils, making these widely known among both singularly famous musicians and ordinary workers in the music industry.

When Schillinger arrived in New York City in 1928, he had little more than the old scores of his Soviet-era compositions. When he died twelve years later, he was a resident of a prestigious building on Fifth Avenue overlooking the Central Park. What lies in between is his extraordinary success as a mentor to creators of American popular music. Schillinger gave private lessons and sold correspondence courses, gradually approaching the status and the wealth of some of the most well-known musicians of his era, and working a punishing schedule. Schillinger's list of pupils includes the Who's Who of 1930s and 1940s American popular music establishment: The composer George Gershwin, Glenn Miller the popular band leader and creator of the evergreen hit "Moonlight serenade" (which was seemingly developed from an exercise Schillinger gave to Miller), and the equally successful jazz band leader Benny Goodman.[6]

Schillinger's sometimes exotic ideas about musical aesthetics and the mathematical ways to "correctly" construct musical material were in very high demand among those who wanted to become or stay successful in the developing American music entertainment business. So much so that even his death did not put an immediate end to their distribution. The posthumous publication of his courses in 1946 and the formation of what later became the famous Berklee College of Music (originally called Schillinger House) carried his influence well into the 1960s. Even Quincy Jones, the music business star who produced Michael Jackson's hit album *Thriller*, has studied Schillinger methods as a student at Berklee.[7]

This line of historical connection not only saved the legacy of early Soviet mathematical avant-garde aesthetics from destruction at the hands of the Stalinist culture of traditionalism and nationalism. It also made available to American musicians many abstract and mathematical methods that were prominent in Soviet arts education and research.

Schillinger created for his American pupils an adapted, but, interestingly, sometimes even more radical version of mathematical aesthetics than the one he brought with him from the Soviet Union. A surprising aspect of his legacy is that his abstract methods were more renowned among creators of popular music than among most of the "serious" American composers of his time, despite popular music's need to be broadly accessible. Indeed, Schillinger's openly anti-traditionalist stance might seem too esoteric to be appreciated by the commercial musicians of his time.

Yet, Schillinger, like many in the early Soviet avant-garde, believed that new musical methods might actually be *more* appropriate for a new public and style because the circumstances of the listener's lives also changed since the time when traditional music theory was codified. Also, popular music of the 1930s and 1940s did indeed differ from earlier styles, and there might have been hope to find in Schillinger some fresh and novel musical-theoretical devices. In addition, the promise of the Schillinger method was

a 3:2

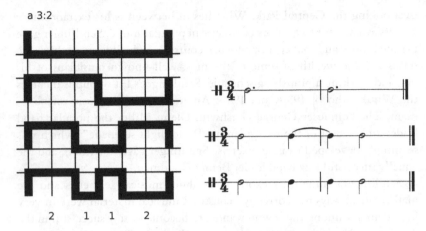

2   1   1   2

FIGURE 6.1 An illustration of the combination of rhythms according to the method taught by Joseph Schillinger.

that it might enable a faster, more "rational," and even almost industrial, workflow, as compared to presumably less effective "intuitive" approaches.

With his seemingly boundless creative energy, Schillinger created innumerable mathematical tricks for coming up with melodies, chords, and rhythms. The most basic of them is the translation of the old idea of ratios into the realm of musical rhythm, akin to the early Soviet approaches described above. The very first lesson for all his American pupils was the demonstration of how to overlay different rhythms so that their pulses would create a ratio such as, for example, 3:2. This means that during the time when one rhythm contains three regular pulses, the other rhythm contains two.

Figure 6.1, derived from a sketch contained in a notebook by Lawrence Berk, shows the Schillinger method for overlaying the rhythms.[8] Lawrence Berk, who later became the founder of the Berklee College of Music, studied with Schillinger in his early years and later used Schillinger techniques in teaching. On the left, the grid contains, from top to bottom, (1) a line showing the full length of the rhythmical cycle (6 units), (2) two elements of three units each, (3) three elements of two units each, (4) the combined rhythm which contains the onsets of both the three-unit and two-unit elements. On the right, the same rhythm is shown in musical notation. In the top-left corner, a small inscription makes clear that the elements form a 3:2 ratio. The rhythmical onsets in the diagram on the left are drawn as vertical lines that go up and down when the line is read from left to right.

This way of depicting rhythm is probably inspired by the movements of a conductor's arm.

While this overlay with a triple rhythm is in fact a well-known figure in Western music, Schillinger only used it as a platform to launch his students onto a seemingly never-ending journey of musical–mathematical strategies. Common ratios such as 3:2 were gradually extended into unusual ones like 16:15, for which Schillinger even used a special demonstration machine, the Rhythmicon (now preserved at the Smithsonian Institution in Washington D.C.). Combinatorics in the style of Mersenne and Kircher (see the chapter on the early modern period) were brought into play, enabling Schillinger to fill pages upon pages of his books with results of combinatorial exercises. New, exotic musical terminology was created in the process, in an attempt to dislodge the centuries-old traditions and foreshadowing the language innovations of the later postwar musical avant-garde.

All this was done by Schillinger in adherence to his firm belief that a more mathematical theory would also be more natural because nature, in his worldview, was intrinsically mathematical. Ironically, his musical writings were littered with countless miscalculations and misunderstood mathematical concepts. And still, they have served as a source of musical ideas for generations of musicians, showing that a helpful musical theory does not necessarily have to be philosophically, mathematically, or acoustically solid. It just has to inspire.[9]

By teaching musicians who worked in popular genres, Schillinger created a now almost unimaginable hidden connection between Soviet arts research and American music industry. It was a path that has led from the bold mathematical experiments that happened during the early years of the Soviet rule to the soothing sounds of an American 1930s big band radio transmission. Buried under the wreckage of once popular musical styles, this path has become all but invisible. However, it is well worth excavating in order to illuminate more fully the intellectual history of popular music, especially now that the computerization of music technology has put mathematical approaches center stage.

In the end, it was an intricate interplay of art and science around the world where the mathematics-inspired movements in the Soviet avant-garde drew on the "objective" ideas about musical aesthetics that came from German 19th-century acoustics research such as Helmholtz. Schillinger, then, finally brought his own version of this mindset into American popular music of the interwar period.

Schillinger's early death prevented him from fulfilling his goal of creating a fully automated system that would have not only included randomization and playback (like in the case of the 19th-century Componium) but also the continuous transmission, via radio, of the mathematical composition.[10] Certainly, he would have made use of the new tool of the digital computer to program his methods which, during his lifetime, were greatly complicated by the need to do the calculations manually. Others have followed in his path, contributing to what became the boom years of mathematical music in the second half of the 20th century.

Notes

1 Ferruccio Busoni, *Entwurf einer neuen Aesthetik der Tonkunst* (Berlin: Berliner Musikalien Druckerei, 1907), https://busoni-nachlass.org/edition/essays/ E010004/D0200001, 30–32; David Nicholls, "Brave New Worlds: Experimentalism between the Wars," in *The Cambridge History of Twentieth-Century Music*, ed. Nicholas Cook and Anthony Pople (Cambridge: Cambridge University Press, 2004), 215.

2 Leonid L. Sabaneev, "Zolotoe sečenie v prirode, v iskusstve i v žizni čeloveka (1959)," in *Vospominanija o Rossii* (Moscow: Klassika-XXI, 2004).

3 Ellon D. Carpenter, "The Contributions of Taneev, Catoire, Conus, Garbuzov, Mazel, and Tiulin," in *Russian Theoretical Thought in Music*, ed. Gordon D. McQuere (Ann Arbor, MI: UMI Research Press, 1983), 293–313.

4 Nikita Braguinski, "Musofun. Joseph Schillinger's Musical Game between American Music, the Soviet Avant-Garde, and Combinatorics," *American Music* 38, no. 1 (Spring 2020): 72.

5 Ellon D. Carpenter, "Russian Music Theory. A Conspectus," in *Russian Theoretical Thought in Music*, ed. Gordon D. McQuere (Ann Arbor, MI: UMI Research Press, 1983), 46. Olga Panteleeva, "How Soviet Musicology Became Marxist," *The Slavonic and East European Review* 97, no. 1 (2019): 100; Emmerich Kelih, "Quantitative Verfahren in der russischen Literaturwissenschaft der 1920er und 1930er Jahre," in *Quantitative Ansätze in den Literatur- und Geisteswissenschaften* (De Gruyter, 2018), 269–288.

6 Nikita Braguinski, "Musofun. Joseph Schillinger's Musical Game between American Music, the Soviet Avant-Garde, and Combinatorics," *American Music* 38, no. 1 (Spring 2020); 57; Alla Bretanickaja, ed., *Dve žizni Iosifa Šillingera. Žizn' pervaja. Rossija. Žizn' vtoraja. Amerika* (Moskva: Moskovskaja konservatorija, 2015); Warren Brodsky, "Joseph Schillinger (1895–1943): Music Science Promethean," *American Music* 21, no. 1 (2003): 45–73.

7 Ed Hazell, *Berklee: The First Fifty Years*, ed. Lee Eliot Berk (Berklee Press Publications, 1995): 190.

8 The drawing in the notebook by Lawrence Berk is contained in: Berklee College of Music. Berklee Archives. Lawrence Berk papers on the Schillinger System. BCA-007. Series 1. https://archives.berklee.edu/bca-007-series-01-theories#page/2/mode/2up

9 I first heard this idea from David Forrest during a meeting at the Harvard University Department of Music in 2019.

10 Alla Bretanickaja, *Dve žizni Iosifa Šillingera*, 247–248.

7
SINCE 1950

From the 1950s onwards, the previously slowly rising topic of mathematical music has erupted in a violent volcano, pouring forth several currents of intellectual inquiry at once, and providing adventurous musicians with a constant onslaught of new, exciting tools.

The need for technological superiority during World War II has led to the creation of many new ideas and approaches whose further development and refinement would become the dominant theme of the following decades. Certainly one of the most visible advances was the gradual creation of the digital computer during the 1940s. At first too costly to be used for anything but military tasks, by the 1950s, the computer became accessible to large non-military institutions such as universities, where musically interested academics were quick to seize it for their cultural experiments.

At the same time, non-technological experiments with randomness and unpredictability gained traction in art music of the 1950s, preparing the ground for their later computerization. In the United States, the star avant-garde composer John Cage drew on the ancient Chinese combinatorial divination text *I Ching* to construct in 1951 his composition *Music of Changes*. In other works, he provocatively employed random physical phenomena such as the scattering of ink on paper to withdraw his own person from the musical outcome and to play around with what he called "indeterminacy."[1] In Western Europe, avant-garde composers such as Karlheinz Stockhausen experimented since the 1950s with *aleatoric* music in which important elements of the composition such as the sequence and overlaying

DOI: 10.4324/9781003229254-9

of musical snippets was left to the spontaneous discretion of the performer or to chance.[2]

At first, the computer was used merely for playback experiments. In a sense, it functioned as a (very expensive) generator of frequencies, with the inbuilt additional capacity to store information about the notes to play. This made it a 20th-century version of Kircher's pinned barrel. Among the earliest examples of such playful appropriation of room-sized electronic "brains" were musical tests in the early 1950s by Alan Turing the British computer pioneer and by the programmers of the Australian computer CSIRAC.[3]

Quickly, however, the true qualities of the digital computer were utilized not only to play, but to generate music. Because it was able to store and run an intricate program defining a sequence of formal steps, the computer could also achieve the more complicated goal of finding *new* melodies that somehow adhere to a set of predefined rules.

July 1956 was the month when two complementary sides of the emerging topic of computer-generated music were presented to the American public at once: A tune called *Push-Button Bertha* written in a style of popular music, and the academic music experiment which later became known as the *Illiac Suite*, consisting of both more historical and more modernist parts.

Push-Button Bertha was created by a commercial company in the then-new computer business, obviously in an attempt to publicize their capabilities. It was a very simple melody generated through a very involved process which included the following:

1. The analysis (by the programmers) of common formal characteristics of melodies in the top hundred pop songs of one year.
2. Translation of these musical characteristics into rules that operate with numbers from zero to nine (representing a scale typical of this musical style) and with logical decisions such as asking if the generated sequence of notes satisfies the programmers' requirements (e.g., they decided, on the basis of the given pop songs, that the tune cannot start on certain notes).
3. Generation of a random sequence of notes from which all notes not fitting the scheme are excluded.[4]

The authors then asked a professional to write lyrics, thus creating a surprisingly competent example of a tune in a popular style. Although some of the rules by which the sequence of notes was generated were discussed by one of the programmers in a publication, the amount of "manual" compositional work that went into the creation of the tune (beyond formulating the rules) remains somewhat unclear.

For example, the article does not explain the creation of rhythm, and the tune contains triple notes for which there seems to be no symbol in the table of possible program output provided in the article. And even if the creation of rhythm was in fact fully automated, it can be presumed with certainty that the programmers used their musical knowledge to choose, among a number of generated sequences, the one that they deemed most impressive, again limiting the role of the program itself. This blurring of boundaries with the goal of presenting the program as more powerful than it was in reality and thus advertising the computer as a general-purpose tool was destined to become a feature of a lot of work in computer music in the following decades.

Illiac Suite, the university computer music experiment, was more ambitious in both scope (a four-movement string quartet instead of a single one-page melody) and sophistication. It involved the imitation of two very different styles and was followed by the publication of a substantial monograph discussing every detail of the composition process as well as many historical forerunners.[5] Especially notable was the use in the *Illiac Suite* of Markov Chains, a mathematical tool based on the calculation of probability of some elements following others (see the previous chapter).

Markov chains have fascinated computer pioneers from the start. In a foundational text of information theory (which has formed the basis for the development of computer science), the researcher Claude Shannon has applied in 1948 Markov processes to the imitation of English, based on the probabilities of transitions between letters and words. At least one example of the outcomes of his experiments deserves quotation as it provides a vivid parallel to how musical Markov chains operate: "IN NO IST LAT WHEY CRATICT FROURE BIRS GROCID PONDENOME OF DEMONSTURES OF THE REPTAGIN IS REGOACTIONA OF CRE."[6]

Here, Shannon calculated the probability of a letter occurring following the given two preceding letters. It is remarkable how close one can get to imitating the sound of a human language (not its more intricate structures, however) with such a simple mathematical procedure. From there, it was only a small step for Shannon to show that human communication is highly redundant, meaning that large parts of information transmitted between individuals can in principle be inferred and even imitated with statistical means. For musical mathematics, this insight meant that imitating human composition processes with the help of probabilities was at least worth a try. Hiller and Isaacson, the authors of the *Illiac Suite* showed that it did in fact work, especially in modernist aesthetics where most listeners did not have a very clear picture of what to expect musically.

At the same time when *Push-Button Bertha* and the *Illiac Suite* made the headlines in the United States, in France the architect-turned-composer Iannis Xenakis tried to put the calculation of probabilities at the very center of his own compositional process. At first, he did not have access to a computer, and tried out his method in a manually created composition, but five years later, in 1962, he was able to collaborate with IBM France. The outcome was his work entitled *ST/10–1 080262.*

Its very name provocatively underscores the technological origin of the composition and denies the listener any real-world references that could contaminate its purely abstract character. Instead, Xenakis imitates the language of engineers by turning the title of his work into a timestamp (February 8, 1962), making it appear as just one in a series of technical experiments. The letters "ST" indicate Xenakis's use of *stochastics*, a mathematical discipline concerned with random processes.[7]

At the heart of Xenakis's stochastic experiments was the idea that novel musical results could be achieved by determining the overall shape of the musical gesture and leaving the generation of the myriads of individual notes that comprise it to the computer.[8] It is a strategy that I like to compare to the use of spray paint in visual art. The person holding the spray can is not trying (and is not able) to control the movement of each and every drop of paint. Still, foreseeable results are achieved because the nozzle of the can sends all drops in more or less the same direction, with areas on the outside being less probable than the ones in the center of the cone of paint.

In *ST/10–1 080262* Xenakis used random musical events in much the same way, shaping the overall outcome by mathematical procedures that ensured that certain events were more probable than others, and thus giving his work a clear musical structure that listeners could relate to emotionally. He also assembled musical gestures and the movements of his piece into a meaningful dramaturgy and made various other compositional decisions during the intricate process of turning columns of numbers which were the outcome of calculations into a score playable by an ensemble.

Interestingly, Xenakis the engineer was always defied by Xenakis the artist in the sense that the clarity of his technical and mathematical descriptions (which would have ensured the repeatability of his experiments) were severely diminished by very artistic language, as well as by what seems like a general disregard of the fact that musically interested readers were not, as a rule, professional mathematicians. An example of this is Xenakis's description of *ST/10–1 080262* in his book *Formalized Music.*[9] Its lack of an attempt to explain in broadly understandable terms the reasons for using the exact mathematical formulas he chose (as opposed to different, or simpler, alternatives) is truly remarkable.

By 1968, the field of computer art, music, and literature had grown, as both the understanding of its promises became more widespread, and the cost of actually using a calculating machine has dropped since the 1940s (but not yet to a point where everybody would be able to do so—this happened several decades later). In this year, an exhibition in London entitled *Cybernetic Serendipity* and its accompanying book showcased the emerging computer artists and their discourse. The word "cybernetics" from which the title is derived is the name of a discipline concerned with processes of control in the broadest sense. Cybernetics is a forerunner of today's computer science, but it also went beyond it by trying to formulate general principles that would be valid in various settings, including not only technological systems but also living organisms and societies.

Besides these computer-based attempts to formalize composition there was also a second, rather non-technical sense in which music was undergoing mathematization during this time. New numerical theories of musical language emerged and became increasingly popular, especially in US-American academia. Among them, the approach with the complicated name *pitch-class-set theory* was probably the most successful—despite, and maybe even because of its mathematically sounding terminology.

It is comprised of several layers of notions which start at the lower end with pitches, continue with arrangements of pitches into sets (without specifying the order in which they could appear in a real piece), and end with mathematical operations such as multiplication.[10] This theoretical apparatus has blossomed in the second half of the 20th century, quickly overgrowing its historical roots such as Schoenberg's twelve-tone technique (see the previous chapter). Its central claim is to be helpful in both analysis and creation of compositions, especially those employing modernist musical language, and its ongoing career is certainly impressive as a sign of how much mathematics theorists of music are prepared to take upon themselves to make their discipline more scientific.

The last decades of the 20th century marked the dissemination of computer-based approaches to music from the confined spaces of well-funded institutions into the everyday settings of modest business enterprises and even the private home. Over time, computing machines gradually became smaller, less electricity-hungry, easier to program, cheaper, and more powerful. For institutions, this meant increasing computing capabilities.

For individuals, this meant first-time access to a *personal* computer, one not belonging to someone else.

This even went as far as providing, from the 1970s on, the children (in the rich regions of the world) with devices based on computer technology and meant for entertainment—the *video game console* and the *home computer*. These electronic machines, though rather cheap and limited in comparison to more professional systems of their time, and certainly primitive in light of later developments, were nevertheless real examples of number-crunching and information-storing computer technology, capable of carrying out quite sophisticated musical algorithms.

When the musician and programmer Peter Langston was hired in 1982 to form a group of people working in the then still-new field of video games, he could already profit from a long history of not just mathematical music, but even true computer music. At that time, the technological challenge of creating a musical background for video games was that while computing power was already somewhat abundant, storage was not. Memory chips and storage media were still too expensive and cumbersome in usage to contain prerecorded soundtracks. Unlike the video games of the later decades these early games had to *generate* music during gameplay if they did not want to annoy the player with endless repetition of the same very short snippets of audio they could store.

Two strategies were employed simultaneously by the authors of early games to make use of what they had at their disposal—sufficiently good calculating processors and some minimal memory resources. First, sounds of music were not played back (as is the case with a digitized recording like an mp3 file) but synthesized note for note (as if the program would press keys of a synthesizer, following a score). This way much less storage was needed because simple instructions such as when to play which note take less space than a digital recording. And second, even the score to be played could be generated and was often at least modified during gameplay.

Langston's team used both approaches when they programmed, first as a kind of a test, the game *Ballblazer*. Notable from the point of view of the history of computer music is how naturally Langston (who did the music part) drew on various procedures of mathematical music in this rather small project.[11] This shows not just his talent, but also how well developed the whole field was already by the early 1980s.

For the historian, the most interesting part of *Ballblazer* is its title tune (the music played before the game starts, likely to be overlooked by anybody not interested in it specifically). The program puts it together from short snippets, like it was the case throughout the centuries-long development of combinatorial music. However, because Langston could make use of a computer, he could integrate into this process additional computational steps that made the output more varied, and he could do it in real time, during playback.

Thus, the choice of snippets was not completely random in this case. The program did first choose several snippets at random, but it then only used the one whose first note was the closest to the last of the preceding snippet. This purely computational trick was a clever way to emulate the melodic continuity stemming in human performances from both the physiology of the hand playing the instrument and the musical tradition.

In addition, Langston made the program leave out certain notes according to a set of probabilities that he defined. By doing so, he emulated, on the one hand, the rhythmical structure of Western music in which certain notes are more likely to be stressed (and thus less likely to be left out) than others. On the other hand, he imitated, through a second set of probabilities, the overarching dramaturgy of a human performer's soloing whose "tension" and thus the density of notes, changes over time. The result was a highly dynamic piece of light jazz in a modern style, absolutely suitable to serve as kind of a pleasant musical wallpaper that it needed to be.

The music-generating routines of *Ballblazer* are a good example of what people generally meant by Artificial Intelligence before today's approaches based on machine learning became widespread. The dominant way to make something artificially "intelligent" was then to program a so-called *expert system*. Expert systems were pieces of software that tried to achieve success in automation by employing data created by experts and processed through hand-crafted algorithms (as opposed to the automatic distillation of knowledge in machine learning).[12]

A simple example would be a program attempting to diagnose an ill person according to knowledge provided by medical experts and using a logical decision-making structure created by a programmer. In the case of *Ballblazer*, Langston (who was himself also a musician) and his friends played the role of the expert by providing snippets suitable for the imitation of the style Langston envisioned, and he also was the programmer who transferred this knowledge into formal instructions that could be acted out by the electronic system.

Langston remained largely unknown by the broader public, but two musicians became the true stars of computer music in the 1990s: the classically oriented David Cope and the avant-gardist of popular styles Brian Eno.

Eno released in 1996 a software album called *Generative Music 1*. As the title suggests, it was a piece of software, with musical parameters provided by Eno, that *generated* the album itself. It made use of a specialized program for creating semi-randomized music called *Koan*, and it created a slightly different version of the album each time the listener started the program. Thus, by buying *Generative Music 1* (and the expensive sound card that it relied on) the listener, interestingly, acquired not the music itself (which at that moment only existed as a possibility), but the means by which it could be produced. Eno's many "generative" projects which were mainly concerned with the formulation of the process by which the piece emerges (instead of the piece itself), as well as his effective communication of the underlying ideas, earned him a lasting prominence in this genre.

Cope's most famous project from this period is his imitation of the style of historical composers such as Bach and Mozart. In the early 1990s, he did some highly elaborate work formulating rules of automatic analysis of music which were so clear that they could be implemented by him as a computer program. The goal of this analysis was to extract from a collection of works by the same composer the features that remain identifiable across the whole oeuvre and can be used to imitate this style in new compositions. This experiment proved highly successful, as Cope's program was able to generate imitations that were almost believable and certainly comparable to "manually" created exercises written in the style of Bach. A recording of some of Cope's early imitations, played back through a mechanical piano controlled by computer data, has been issued as *Bach by Design* in 1993. It provides a good starting point for a listening journey of 20th-century computer music.

With their genuinely artistic intensions, Eno and Cope went beyond Langston's primarily experimental and practical interests. Yet, as can be seen from the amount of mathematical and computer-based approaches that have accumulated by the end of the 20th century, there remained little really new territory to explore, at least until more powerful computers made machine learning a reality for the average programmer. This following, and still continuing, phase in the history of mathematical music is the subject of the next chapter.

Notes

1 It has been noted that Cage's use of ink dots corresponds to a satirical idea from 18th century. Lejaren Hiller and Leonard Isaacson, *Experimental Music. Composition with an Electric Computer* (New York: McGraw-Hill, 1959), 52, 53; Nick Collins, "Origins of Algorithmic Thinking in Music," in *The Oxford Handbook of Algorithmic Music*, ed. Alex McLean and Roger T. Dean (Oxford: Oxford University Press, 2018), 69.

2 Klaus Ebbeke, "Aleatorik," MGG Online, 1994.

3 Paul Doornbusch, *The Music of CSIRAC. Australia's First Computer Music* (Australia: Common Ground Publishing, 2005); "First Recording of Computer-Generated Music – Created by Alan Turing – Restored," *The Guardian*, September 26, 2016. https://www.theguardian.com/science/2016/sep/26/first-recording-computer-generated-music-created-alan-turing-restored-enigma-code

4 Martin L. Klein, "Syncopation by Automation," *Radio-Electronics*, June 1957: 36–38; Alex Di Nunzio, "Push Button Bertha," Musica Informatica, 2013. http://www.musicainformatica.org/topics/push-button-bertha.php; Matthew Guerrieri, "'Automation Divine'. Early Computer Music and the Selling of the Cold War," NewMusicBox, October 10, 2018. https://nmbx.newmusicusa.org/automation-divine-early-computer-music-and-the-selling-of-the-cold-war/

5 Lejaren Hiller and Leonard Isaacson, *Experimental Music. Composition with an Electric Computer* (New York: McGraw-Hill, 1959); Douglas Keislar, "A Historical View of Computer Music Technology," in *The Oxford Handbook of Computer Music*, ed. Roger T. Dean (Oxford: Oxford University Press, 2009), 33.

6 Claude Shannon, "A Mathematical Theory of Communication," *The Bell System Technical Journal* 27, no. 3 (July 1948): 388, https://doi.org/10.1002/j.1538-7305.1948.tb01338.x

7 Damián Keller and Brian Ferneyhough, "Analysis by Modeling. Xenakis's ST/10-1 080262," *Journal of New Music Research* 33, no. 2 (2004): 162.

8 Peter Hoffmann, "Xenakis, Iannis – 7. Macroscopic Stochastic Music," *Grove Music Online*, 2016.

9 Iannis Xenakis, *Formalized Music. Thought and Mathematics in Composition* (Stuyvesant, NY: Pendragon, 1992), 131–54.

10 Michiel Schuijer, *Analyzing Atonal Music. Pitch-Set Theory and Its Context* (Rochester, NY: University of Rochester Press, 2008); Catherine Nolan, "Music Theory and Mathematics," in *The Cambridge History of Western Music Theory*, ed. Thomas Christensen (Cambridge: Cambridge University Press, 2006), 289–95.

11 Peter Langston, "Six Techniques for Algorithmic Music Composition. A Paper for the 15th International Computer Music Conference (ICMC) in Columbus, Ohio, November 2–5, 1989," 1989. http://www.langston.com/Papers/amc.pdf; Nikita Braguinski, *RANDOM. Die Archäologie der elektronischen Spielzeugklänge* (Bochum: Projekt Verlag, 2018), 52–60.

12 Rebecca E. Skinner, "Artificial Intelligence," in *Debugging Game History. A Critical Lexicon*, ed. Henry Lowood and Raiford Guins (Cambridge, MA: The MIT Press, 2016), 29–36.

… to possibilities

8
POWERFUL AND LIMITED
(INTRODUCTION)

When discussing music AI, or AI in general, it is important to keep in mind what is real about it at this specific moment in history, and what belongs into the realm of possibilities, including utopian hopes, dystopian fears, science fiction, or even corporate public relations.

As computer technology grew over the last decades, so, too, changed the notion of artificial intelligence. Some of the computational methods that were considered "AI" even a couple of decades ago have largely lost this status due to the development of new, more powerful tools.

Today, the dominant and defining technology that is deemed worthy of the precious "AI" label is *machine learning*, and, specifically, its modern variety called *deep learning*.

In deep learning, incoming information is fed to a large group of very simple calculating units (the so-called neural network), and the results of these calculations are used to either segment this information in a meaningful way, or to generate new information that somehow resembles the original.

In music, it can be, for example, a set of snippets of notation from Mozart as well as from the modernist composer Arnold Schoenberg, which the machine learning system sorts into two groups according to their style. Or it can be a new piece in the style of Mozart (or Schoenberg), generated on the basis of the provided examples.

Before a deep learning system can do its work, it normally has to be "trained." This means that initial information (its *input*) is provided,

DOI: 10.4324/9781003229254-11

together with the correct *output* for each given input. Then, during training, the system adjusts the parameters of the calculations carried out by the individual units of the network so that their overall result approximates the correct output.

In our case, it means that the snippets with Mozart's music will be fed to the system together with the label "Mozart" for training, and the calculating parameters will be automatically adjusted until the system starts to output the label "Mozart" when given a Mozart piece, even if it was not among the examples used to train the network.

This procedure is both surprisingly powerful and astonishingly limited.

On the one hand, it seems capable of imitating the results of one of the last strongholds of human thinking - our intuition. Nobody is surprised anymore if a digital calculator does arithmetic calculations more quickly and more reliably than a human. The mechanical execution of rigid mathematical rules, and the storage of clearly defined information, are the digital computer's forte.

Yet, as of this writing, it is still a small wonder that a deep learning system can do some things that humans do intuitively, without conscious calculation, and often even without being able to explain how they arrived at the conclusion.

People who actively work in the field of deep learning sometimes like to contrast it to human intuition, stressing how the algorithmic number-crunching delivers data-driven decisions whereas intuition is not grounded in data and objective calculations. In contrast to this view, I prefer to see deep learning and intuition as two very similar procedures, as they both involve learning to make a decision without creating an intermediate theory that is separate from the procedure itself. To ponder the parallels between intuition and deep learning, you can consider a small child learning to sing. It does not operate with abstract concepts from music theory, but imitates, through (playful) trial and error, the singing that is found in its environment.

Mostly, there seems to be no clear, formalizable, rule-like parameter that we, humans, consciously rely upon when we make intuitive decisions, such as, for example, when we tell whether the vowel we hear in speech is an "a" or an "o." And yet, the algorithmic tool of deep learning can be very effective in many such cases, including the transcription of speech.

This is of course only possible because vowels do objectively differ in their acoustic properties, and systems can therefore learn to replicate the intuitive decisions made by humans on the basis of these properties. In a sense, today's deep learning systems are the technological outgrowth of the old idea of the *unconsciously* calculating mind, or "soul," as Leibniz put it in the 18th century (see the chapter on the early modern period).

The flip side of this new power of AI is that a large amount of training data is normally needed to adjust the calculating parameters of the deep learning network before it starts to work reliably. In our Mozart AI example, somebody would probably have to manually label a large number of music snippets, which is a considerable investment of human mental work. Also, the programmer would need to make multiple decisions regarding the internal structure of the network, as many different possibilities have been suggested in deep learning research, and there is no one approach that fits all situations.

In the end, it might become easier and cheaper to not train the network at all, and to just ask the worker to tell whether a piece is from Mozart or Schoenberg each time such a question might arise. In all likelihood, as the result of working on this task, our worker would also produce less climate-endangering waste than the computer.

Still, in contrast to our fictional example, there are of course many situations in computer music in which deep learning is a meaningful and efficient tool. To understand its strengths and limitations, let's take a look at how it functions on a more fundamental level.

9

HOW DOES DEEP LEARNING WORK?

Deep learning is a sub-field of *machine* learning. This second, more general term describes all the methods of letting a computer system semi-automatically derive *features* from the input that the user wants to become its *output*.

To illustrate this process of deriving features, let me introduce another fictional example. Suppose that someone who does not have knowledge of Western music theory wants to build a computer system that only outputs chords that this person finds "pleasant." This person has been exposed to Western classical music since childhood and has a firm intuitive feeling whether a chord sounds "pleasant" (in which case it is very probably a chord commonly used in the music of this period) or "unpleasant" (mostly because it is unusual, and has not been engrained through centuries-long use). Music theory provides definitions for chords employed in Western music, such as the major or the minor chord. It defines these as combinations of three different notes with specific distances (intervals) between the notes. Yet, this person knows nothing about music theory beyond the idea that a piano keyboard has keys each of which plays a specific tone, and that chords consist of several notes.

To achieve its goal of making a machine that plays only "pleasant" chords, the person then creates a large table in which it writes down many random combinations of notes, together with the information whether this particular combination sounds "pleasant" or not. In the next step, this person builds a computer system that takes the table as its input and learns to infer the hidden relationship between the structure of the combination and

DOI: 10.4324/9781003229254-12

the label that has been assigned to it. With each new example, the system adjusts itself slightly to minimize its error in predicting the label given by the person. In the end, the system becomes capable of predicting with some certainty which combinations are likely to be perceived by the person as "pleasant."

What happens here is that the *machine learning* component in the system learns to derive the needed feature ("pleasant" or not) from the input (the combinations of notes and their ratings). It therefore *learns the relationship between the input and the output*. Once the person has decided that the system has learned the relationship with sufficient precision, the training phase is complete, and the system can be used to replace human work in whichever area of activity it was designed to replicate or augment.

Today, the most visible advances in music AI happen in the field of *deep* learning. To understand how deep learning differs technically from other forms of machine learning, let's look at how the calculations that enable the system are organized.

The basic calculating units that underpin the modern approaches to AI work in the following way: They receive several numerical values as their inputs, do calculations with them, and output a numerical value. The parameters of these calculations are adjusted for each unit during training until the system starts to give the needed overall output.[1]

For historical reasons, and because this type of machine learning was initially inspired by how the brain is organized on the biological level, the calculating units are commonly called *neurons*, like the nerve cells. And because there are normally many neurons in a system, with each one connected to several others, these computer systems have received the name of *neural networks*.

In principle, it is possible to do some very basic work with just one neuron. Historically, this was also what scientists experimented with first during the early decades of AI research. Yet, to let the system learn more complicated relationships between inputs and outputs, it is helpful to use more neurons, and to organize them in *layers*.

Figure 9.1 shows an example of a network with four layers. On the left, two neurons make up the *input layer* of the network. They provide the data on which the system does its calculations. The arrows indicate how the outcome of one neuron's calculations becomes the basis for other neurons' work.

On the right, one neuron serves as the final *output layer*. This is the place where the user of the system receives the result. In the middle, this example has two layers of additional neurons that enable the network to learn a more complex relationship between provided data and the desired output.

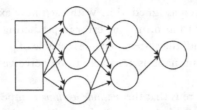

FIGURE 9.1   A neural network with four layers of neurons.

The layers positioned between the input and the output layers are called *hidden layers*. If a network has a minimum of two such hidden layers, it is called a *deep* learning network.

Most of the computer science theory that is needed to create a deep learning network has already existed by the end of the 1990s.[2] However, it took almost fifteen additional years for deep learning to really gain strength, start to outperform other approaches and capture the public imagination. What was it that happened during this period and made deep learning the star of the artificial intelligence scene?

Two factors have enabled its explosive growth: the emergence of big collections of data and the lucky idea that computer hardware that was originally created for a completely different area of use can be repurposed to speed up the network's calculations.

"Big data" is the term commonly used to describe the situation in which most of us live by now where almost everything people do online leaves a trace of data that, combined, gives enough fodder to train effective networks. Add to this the flood of data from automatic sensors (such as weather stations, for example), as well as the heaps of documents that populate the computer systems of big companies and institutions, and it becomes evident that one of the two prerequisites for training a network—the abundance of examples—is met.

Yet, the second prerequisite—cheap calculating power—was also unavailable until the mid-2000s. A trick solved the problem: Computer parts originally created to enable more realistic visual effects can be used to do the number-crunching in neural networks. Graphical processing units (GPUs) were first developed and mass-produced at relatively low cost as an add-on (the so-called graphics card) that users could buy to outfit their computer for modern video games. By 2004, researchers realized that GPUs are capable of doing exactly the same kind of calculation that is needed for neural networks, and by 2007 a well-known manufacturer made it possible for programmers to easily integrate GPUs into their AI projects, opening an entirely new field for its hardware.[3]

All of this can seem like a self-sufficient miraculous machine where everyone can just push a button (provided, one has the right hardware), and a solution will appear, no matter what question needs to be resolved.

The reality is, however, different. The value of the neural network often comes from human work that enables its creation.

Large datasets that are interesting targets for training such as, for example, images of vehicles and specific road situations often require a lot of manual input which only seems to be small to each contributing person if it is distributed among a very big crowd. The cumulated effort is enormously large, and it can become prohibitively expensive if well-paid experts are needed to build the dataset.

At the same time, a lot of human labor is hidden inside the technical structure of each system, and I do not even mean the decades of computer science research here, but the labor each programmer needs to invest to make numerous decisions about which type and configuration of machine learning to employ in each specific situation.[4] Even in those cases where the process of finding a working configuration is replaced by a series of automatic trial and error attempts, the programmer needs to make educated guesses and employ professional knowledge at all stages of the project.

The palette of potential technical solutions is almost endless,[5] and needs to be navigated with care. So far, we have only considered neural networks which receive during training the correct answer together with each initial input. This enables them to learn the hidden relationship between input and desired output. However, it is also possible to use some of the techniques of machine learning without providing the correct results up front. This is called *unsupervised* learning. Here, the system tries to segment incoming information into meaningful groups according to similarity, without the operator telling the system which kind of similarity is the most important.[6]

One can also let two networks compete with each other. In this scenario, the first network attempts to generate data that is as similar to the provided dataset as possible, and the second network tries to distinguish the original data from the fake data. Both networks are automatically adjusted to react to the other network's actions until the generating one becomes sufficiently good at faking the data. This procedure goes by the name of generative adversarial networks, or GANs.[7]

It is even possible to try to simulate the behavior of a simple robot (called "agent") acting inside a controlled environment with the goal of letting the robot learn the most efficient course of action through trial and error, coupled with computational "rewards" that tell it whether an experiment was successful. This kind of machine learning is called *reinforcement* learning.

The list of possibilities grows daily as researchers at both academic institutions and big commercial companies continue to come up with new, even more powerful and efficient ideas, while at the same time competing with each other for the best performance in machine learning tests.

The data that is seemingly everywhere nowadays in reality also needs a lot of work before it can be used for machine learning. Someone has to collect it in the first place and then to filter it, making important decisions which features to use and which ones to discard for each specific project.

In short, next time a newspaper article or a blog post indiscriminately conjures up a picture of a miraculously self-working "AI," imagine all the hard work that really makes possible its creation and functioning.

At the same time, there are also current developments that promise to turn machine learning into a much more approachable technology.

As mentioned above, in supervised learning a model needs to be trained with initial data before it can do the calculations required for the segmentation or generation of new data. However, this training does not need to happen at once. It is perfectly possible to train a model only partially, and then to add on top of this training several different additional trainings, leading to the creation of independent models, each specifically suited for its own task. This way, a lot of training work can be reused. Depending on how big and powerful the partially pre-trained model is, the amount of additional training may be quite small in some cases.

One example of such a very large and powerful model is GPT-3 by OpenAI, a system from 2020 that generates imitations of written language. It is in fact so big that it can fulfil many language-related tasks with very little or even no additional training. The authors of GPT-3 also made available a simple interface where the user can provide, without resorting to programming, some absolutely minimal context for the envisioned task, which is mostly already enough to steer the output of the system in the needed direction.

This procedure is so simple that it does not even seem to deserve the name of additional training, or fine-tuning, as the programmers call it.

Instead, the authors of the system speak of "few-shot," "one-shot," and "zero-shot" learning by which they mean the situations where a handful of additional examples, just one example, or even no example at all are provided to orient the system in the desired way.

This is possible because GPT-3 is trained on a big part of everything that is available online in written form. And this is such an enormous collection of human knowledge, communication, artistic output, and all else that the model seems pretty much pre-trained for most of the tasks that a typical user could come up with. Even some simple musical tasks seem to lie within the reach of this model. Giving a little bit of context and making a couple of simplified adjustments such as how random the output needs to be are often enough. Here are some examples.

1. Chords

Using standard settings and the numbered structure recommended by its authors, I provided the system with the following prompt:

This is a music harmony generator that creates a sequence of chord symbols.
1: C Am F G C
2: Dm Gm A7 Dm
3: G C6 D7 G
4:

The output of the system was:

```
  F G A C F
5: Dm Am7 G6 Dm
6: G C F A C7
7: F Dm7 G#dim Dm
8: G Dm7 Gm A
9: F G6 G7 C
10: C Am G C
```

In my prompt, I used keywords that make it clear that I want the system to generate a very specific kind of text, the chord symbols such as C or Am which are a standard part of Western theory and notation, especially in popular music. I also used very common sequences that follow a widely known tradition of Western music theory that recommends the choice and order of chords in a given key. It can be said with certainty that my example sequences and many similar ones appear very frequently on the internet on websites and in books dedicated to music theory, and were therefore a part of the initial data on which GPT-3 was trained.

In the output, it can be seen that the system reacted correctly, providing seven additional sequences of chords, and numbering them accordingly. Only one of them directly repeats one of the initially provided examples (C Am G C). All the others follow the prompt in a more "fuzzy" way, containing four or five chord symbols. All these symbols do exist in music theory, so the system did not make up a chord like Z1 which would only superficially resemble the pattern of a letter with a number visible in the prompt. Some of the generated sequences also more or less follow the traditional harmonic movement which I employed in the prompt while others (like the number seven) can be seen as either more adventurous or more random, depending on the user's point of view.

Given that chord sequences expressed as symbols made from letters and numbers are also a kind of text, this experiment can be seen as a success. I had a "creative" task (generating chord sequences), and I more or less accomplished it with a minimal amount of additional training.

It is true that this task can be solved much more efficiently using much simpler techniques such as randomization in the style of Athanasius Kircher's wooden box from the 17th century (see the chapter on the early modern period). And even machine learning itself can be put to more effective use by training a model not on everything available online, but on sequences of chords and nothing else. This simpler system would also arguably deliver a much more robust and believable imitation of human harmonic thinking. Yet, all these approaches mean additional work, and, compared to a model that is less efficient but already available, it is very well possible that a user will turn to the latter.

2. Guitar tab

In my second experiment, I chose a slightly different version of GPT-3, which is tuned more towards following a written command instead of continuing a list of examples. I gave it the following prompt:

Write some music in guitar tab format.

The result was:

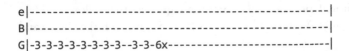

```
e|----------------------------------------------------|
B|----------------------------------------------------|
G|-3-3-3-3-3-3-3-3--3-3-6x-----------------------------|
```

```
-once
D|-3-3-3-3-3-3-3-3--3-3-3x----------------------------|
-twice
A|-1-1-1-1-1-1-1-1--1-1-3x----------------------------|
E|---------------------------------------------------|
```

The guitar tab format is a common way of sharing notes for the guitar on the internet. It uses text symbols to visually present the musical information such as where to place the fingers on the guitar's neck and which rhythms to use. As can be seen in my prompt, I did not give any examples, only the command containing the needed keywords. At first sight, the output is almost credible. It does contain the correct vertical arrangement of six lines corresponding to the individual strings of the guitar. The lines are also all of the same length, correctly indicating the end of the musical phrase. The numbers placed on the strings G, D, and A form a very common figure in rock music, a so-called "power chord." The music lacks change as the same chord is repeated over and over. But, then again, I only asked for "some music," not for "good" or "interesting" music. So far, the output looks almost okay.

However, there are also some incorrectly rendered elements that fall in the (absolutely unscientific, but strangely fitting) category of "AI weirdness." The playing instructions "once" and "twice," which are a common way to tell the musician to repeat the phrase, are inserted here right between the strings and placed above each other, making it evident that the system does not really understand what it does. Or does it? After all, the phrase to be played on the G string is supposed to be repeated six times (it contains the command "6x" at the end) and then only once as a whole, while the one for the D string will be repeated three times, and then twice... It is hard to tell, but I would rather vote for this being a coincidence, especially given that the phrase for the A string is not repeated at all.

To understand where this "weirdness" comes from, it is helpful to look at how the system operates. It predicts which piece of text is likely to appear next, given the beginning, which is the prompt, and the training data, which has been distilled into the parameters of the network. It is very effective at imitating a specific style of writing or conversation. Yet, it does not have a conceptual grasp of the texts it simulates, and it does not employ the human methods of reasoning that originally enabled the creation of data on which the system was trained.

This sometimes leads to funny situations where the system's probabilistic structure suddenly makes itself clearly visible in an unexpected way. I was lucky enough to witness the system in a state of development where this interesting behavior occurred. With more use and more feedback from users, cases like this are likely to be eliminated over time.

But here it is: One day, using the "chat" settings and prompt, I asked the system the following simple question:

`What time is it in Berlin?`

The answer was:

`It is 8 pm.`

At first sight, this seemed like a valid reply. The problem was, however, that it was actually 10:24 am. The system did not "understand" my question in any way comparable to what a human would do in this situation. Instead, it "saw" in its network parameters that a question like that is often followed by an answer like the one given, and it generated a semi-randomized output that best fitted the training data. Asking the same question again would generate another seemingly fitting, but actually absurd response, with a different time.

This is the fundamental difficulty that people working with such systems need to keep in mind while using them. It is all too easy to fall for the seeming humanness of the output with its sometimes almost-perfect imitation of human discourse. The surprise is all the greater, then, when the output is absolutely not what the user expected.

This is all due to the fact that systems such as GPT-3 work with probabilities. When I experimented with the prompt "Classical music has enjoyed a," the system chose the word "renaissance" as the most probable continuation, with an 11.94 per cent likelihood. But it also considered "resurgence" (10.01 per cent), "revival" (8.73 per cent), and many other words which all lead to their own branch in the giant tree of probable continuations. Such an approach is guaranteed to give *some* sort of output, even in those special cases where it is ultimately unusable.

In the next chapter, I will first take a look at the history of mathematical methods in music, from the earliest ones described at the beginning of this book to the most recent examples of machine learning tools that I have discussed above, to try to understand what it really is that connects them. Then, I will also point to some difficulties that can arise when people try to bridge the gap between human musical theory and a deep learning model. Such difficulties suggest that the path from antiquity to AI (which is the topic of this book) may in fact lead through a certain ravine which is challenging to cross (but not impossibly so).

Notes

1 The following paragraphs are grounded in the exposition given in John D. Kelleher, *Deep Learning* (Cambridge, MA: MIT Press, 2019), 67, 68.
2 John D. Kelleher, *Deep Learning*, 143.
3 John D. Kelleher, *Deep Learning*, 153, 154.
4 John D. Kelleher, *Deep Learning*, 23–26.
5 Jean-Pierre Briot, Gaëtan Hadjeres, and François-David Pachet, *Deep Learning Techniques for Music Generation* (Springer, 2020).
6 John D. Kelleher, *Deep Learning*, 27, 28.
7 John D. Kelleher, *Deep Learning*, 235.

10
PUTTING MUSIC AI IN PERSPECTIVE

Looking back at the story told so far, one can ask whether machine learning is at all connected to the earlier attempts at putting mathematical methods to musical use. One could argue, for example, that the sheer number of calculating units in a deep network gives rise to a completely new range of emergent "behaviors" that are absolutely impossible to achieve with smaller-scale systems.

Let me try to address this question. In what follows, I would like to offer two arguments against the idea that music AI is so fundamentally different from earlier mathematical music that it does not make sense to combine the two topics in one book (as I did).

The first one is that without the earlier mathematical attempts at understanding and creating music the very definition of music would be so different as to make any mathematical approach, such as deep learning, seem absurdly misplaced.

And the second one is that machine learning is so deeply rooted in earlier, pre-digital tools of mathematics, including the ones that were used to experiment with music, that any attempt to extract it completely from its history automatically leads to a skewed and unhelpful picture.

It is true that the more complex inner dynamics of deep networks enable a new sort of transformations between input and output. In other words, a simple system based on randomization such as Kircher's 17th-century wooden box, of course, does not even begin to resemble the abilities of, say, GPT-3 to react to inputs written in a natural language and to generate a response.

DOI: 10.4324/9781003229254-13

Yet, without the historical-cultural background that in my view makes them seem possible and desirable in the first place, machine learning systems for music would not have become the active and attractive area of experimentation that they are today.

Deep learning systems do offer new possibilities, especially the automated extraction of typical features from a corpus of music, which were unthinkable in earlier decades, let alone centuries. And still, it was the history of attempts at doing the same with simpler tools that made the idea of using machine learning for music seem at all probable.

This long history of gradual and mutual rapprochement between the discourses and tools of music and mathematics made machine learning look like the logical next step in music, after the ratios, combinatorics, statistics, and everything else that I have described in the first part of this book.

Earlier attempts prepared the ground by showing that music is, indeed, susceptible to mathematical processing. Multiplying or dividing the length of strings by small integers led to tone relations that were accepted by the listeners of previous centuries as "musical." Or, rather, these mathematical procedures defined what was a "musical" tone relation in the first place. The centuries-long build-up of mathematical tools to discuss and create music has slowly redefined the very notion of music itself, step by step bringing it closer to a state where it could be more easily expressed and manipulated in terms of symbols and numerical values.

Dividing the range of all possible frequencies into discrete "compartments" called notes, for example, has for centuries enabled the notation of musical pieces. But it also facilitated the later computerized approach because at this point it was already easy and logical to replace note names with numbers and to do calculations with them. Likewise, randomly shifting around snippets of music has long proved effective in creating more or less credible imitations of human musical creativity. After centuries of such experiments, early computer music often continued in the same vein.

When one starts to look back at history like this, the seemingly abrupt boundary between machine learning and earlier methods becomes more and more blurry the more one considers small intermediary steps. Before researchers started to assemble massively large deep learning models, they conducted "AI" experiments with single calculating units, whose mathematical structure and "behavior" was surprisingly simple. It was only through their decades-long growth in size and complexity that the machine learning models gradually began to acquire the new qualities that now set them apart from earlier attempts.

I would even go as far as to argue that the machine learning approach is, despite the great difference in speed, not *fundamentally* different from similar attempts that were carried out by hand at the beginning of the 20th century. It is true that manual data-collection and calculation made the process absurdly slow and error-prone. But the very possibility of carrying it out already changed the scene. The experiments by the Russian mathematician Markov mentioned above in this book happened decades before the digital computer crossed the boundary between science fiction and reality. And yet, they already distilled cultural data (poems) into a numerical representation through a mathematical procedure. Along the way, the Markov method derived features (probabilities of consonants and vowels following each other) that helped generate imitations of human language.

The big question posed at the beginning of this chapter (can one put ideas from antiquity and AI in the same book, as I did here?) also connects to a long-standing discussion in the study of technology whether it is the cultural background that leads to technical experimentation and change or whether, conversely, new technology determines the pathway for cultural development.

In fact, this is a bit of a chicken-and-egg problem where each side enables further changes in the other. In relation to the interplay of music, mathematics, and technology, this means that, historically, waves upon waves of cultural and conceptual, but also technical change have brought us, through their complex dynamics, to our current state of music.

It does not mean that other outcomes were not possible. It also does not mean that this history followed a perfectly linear path, without any dead ends, breaks, branches, or loops. But it means that, without a knowledge of the past, our ability to understand the current state and to try to predict (and shape) the future of music technology is diminished.

Now, after looking at some of the history that connects musical AI and its predecessors, let's also take a look at what separates them.

One feature of machine learning that is especially interesting in relation to music generation is its seeming ability to produce credible imitations of language, complete with a correctly sounding grammar and a stylistically consistent use of vocabulary. The GPT-3 system that I used as an example in the previous chapter is very effective at writing human-like texts of all sorts, so much so that it could even generate more-or-less working imitations of such exotic text genres as chord progressions and guitar tab notation without any additional training.

Astonishingly, such imitations of language can be created without providing the machine learning system with a grammar stated in any explicit, or open and direct, way.[1] This is contrary to how people normally approach the task of learning a foreign language, or even of learning to use a formally correct version of their own, native language. This is also notably different from earlier approaches in mathematical music, where the author of the system explicitly stated the rules of the game, such as "take exactly these snippets of music I have prepared and don't ever use one of the finishing snippets at the beginning of the piece," etc.

At first, AI researchers attempted to create language-processing and language-generating systems out of the basic building blocks of human grammar rules. Yet, these experiments often delivered very clumsy and inauthentic imitations, and were later outperformed by deep learning as soon as it became popular.

In this context, it is certainly an interesting experiment to try to generate music that sounds *as if* it were based on explicit music-theoretical rules, but is in fact only derived from the work of composers who more or less adhered to these rules.

Music education is notorious for its transfer of a large number of such explicit rules (in addition to the more elusive *implicit* traditions which are not even communicated directly). Everyone who learns to play Western chords on a guitar, to count traditional Indian rhythms, or to play in an Indonesian Gamelan ensemble, needs to understand and internalize some sort of an agreed-upon theoretical framework that enables communication between musicians and coordinated play. Importantly, the stability of this framework also creates a shared ground of broadly understandable musical signs between the musician and the listener.

A good example of an explicit musical rule is the convention, in Western music theory, that a major chord consists of three different notes with specific intervals between them. To create one version of this chord, one can start with any key on a piano keyboard and add to it the notes produced by the fourth and the seventh higher keys, without making a distinction in counting between black and white keys (see Figure 10.1).

FIGURE 10.1 Different major chords on a piano keyboard.

Musicians learning to play music professionally internalize this knowledge (which is normally codified through a much more elaborate network of interconnected theoretical notions). At the same time, listeners learn, through practice, to intuitively recognize the sound of, say, a major chord, or a specific Indian raga, and to associate it with a certain emotion, even without knowing about the musical theory behind it.[2]

People without a background in music theory are not able to convincingly imitate rule-based styles such as, for example, virtuoso piano pieces from the 19th century. And yet, it is, surprisingly, possible to create a machine learning system that produces exactly this kind of music associated with professional musicianship and applied music theory. In the realm of machine learning, all this can be done without providing explicit rules. Instead, the system is capable of deriving everything from examples.

An impressive proof of this ability of machine learning systems is an experiment that was carried out in 2019 by scientists working for the Google Magenta research project. They succeeded in building a machine learning set-up capable of making credible imitations of human piano composition and performance that have a clear, audible similarity to music based on the rules of music theory and played by a skilled pianist.[3]

There are, however, also important barriers that still stand in the way of musicians employing machine learning systems effectively in their projects. A presentation that was given in 2020 by researchers from the same group highlights some of these difficulties.[4] The most serious obstacle is that musicians cannot currently use a machine learning system in the same way as they would have used their own musical abilities.

A person with even minimal exposure to the musical tradition will be intuitively capable of playing around with acoustic material and adjusting it to the kind of emotion or effect it strives to achieve. After just minutes of experimentation with a piano, everyone can come up with the idea that aggressively hammering out deep, bell-like tones on the lower keys creates a dark, menacing atmosphere while slow, mild tinkling in the mid-range is suitable for more idyllic pieces.

Yet, as of this writing, it is very difficult to intuitively adjust the mood of the music generated by a machine learning system in a similar way. This makes current systems cumbersome to use in a real-world musical setting. Sometimes, people who experiment with them even resort to a kind of a "needle in a haystack" method where they automatically generate inordinate amounts of data and sift through them manually, throwing away everything but the one melody that they really needed.

This problem raises an interesting question: Could existing music theory that humans created for themselves also provide useful signposts for the automatic search for patterns in the data that is at the core of machine learning?

On the one hand, listeners are used to hearing the results of the rules and recommendations of music theory in the music that surrounds them, especially in the case of historical styles and light background music whose function is to be soothingly predictable. Ignoring them should, theoretically, render the generated music less human-like. On the other hand, the power of machine learning systems lies exactly in the way they choose for themselves which features they need to derive from the input in order to approximate the desired output. Interfering with this process by imposing on the system some explicit rules might in fact make it less efficient. Researchers will certainly need to address this question (and I am sure that some of them already do).

This question can even be rearranged to point to a fundamental difficulty with deep learning: Do features derived from the musical input by the technical system resemble in any way the musical theory that humans relied on throughout the centuries? In other words, does the machine learning system have its own musical theory, encoded in the parameters of its calculations? The problem is that even if it does, we cannot use it because we are not able to mentally integrate and adopt the myriad calculations that lead, in the deep network, from the input to the output.

The deep networks of today are very difficult to interpret.[5] They give a result, but why or how they arrived at this exact result, often remains completely incomprehensible because of the large number and the interconnected nature of calculations.[6] Adding and multiplying millions of small numbers, and keeping them all in memory is just not the way humans think (at least not consciously). We do know that addition, multiplication, and a couple of other mathematical functions lead directly from input to the desired output. We can even try to see which examples in the training data contributed to the system's answer. But this analysis of the network still does not help us master a task in the same way a good (human) theory book does.

It is even questionable if the *results* given by deep learning systems can teach us anything. In the case of the board game Go, a system called AlphaGo, which includes a machine learning component, was the first to beat a human champion. This led to a broad movement in the Go community of imitating some of the moves played by the program. At the moment, it is very *en vogue* to play "AI" moves, even among beginner players, and to value them more highly than traditional moves based on centuries of human play. Yet, it is perfectly possible that traditional Go knowledge and theory are actually better at facilitating the human thinking process than the mathematically more effective, but inhuman "AI" moves. What is worse, players might be imitating the wrong part of AI, with the move they value so highly being actually an unimportant byproduct of some other hidden process that is really the source of AI's playing strength.

Is the same going to happen in music? Are future aspiring composers going to fall prey to the seeming superiority of some novel and trendy "AI" chord or sound, without understanding why the program generated it in this specific context? Time will tell, but it is notable that such popular fixation on one external detail in the output of the internal thinking process of a (human) composer has already happened in the past: Richard Wagner's "Tristan chord" and Alexander Scriabin's so-called "Mystic chord" both captured the imagination of their time.

Go is a game with stable, generally unchanging rules and a clear numerical result that is perfectly suited for use as a measure of success in training the network. By contrast, there are no unchangeable rules in music, and no objectively valid numerical result that tells how good the music is. Any application of machine learning in a musical setting is therefore an individually crafted, personal solution that seems to work for its creators, and for which its creators hope that it will work for other people, too.

An important distinction in machine learning experiments at generating music is between those that work with notation and those that entail the direct creation of audio. While creating credible notation, or some other kind of simplified information about when to play which note, is already difficult enough, the generation of audio poses even more extreme demands on researchers and their equipment. Yet, such attempts at directly capturing the nuances of musical performance from the recordings that are the training data are very intriguing because they promise to also imitate the subtle details of musicianship that are not expressed in notation.

Working with audio data, instead of data that represents musical notation, requires much more computational resources, arguably removing the resulting machine learning system even further from its small-scale historical predecessors (without, however, breaking the bond completely, as I have argued above).

In 2020, a group of scholars working for the research organization OpenAI published the results of their Jukebox project, which was a very successful and large-scale attempt at direct audio generation in different styles of popular music with the help of machine learning techniques.[7]

The system was trained with songs from various popular genres. It learned three different compact representations of the music from each song, with one representation catching the bigger structure of the pieces, and the two other representation mimicking the smaller details. Using the trained system to generate new examples, researchers were able to create very credible imitations of various styles, with extremely human-sounding and sometimes truly enjoyable musical elements, including expressive singing that even fits the flow of the lyrics in some of its more fortunate moments.

Musicians know that notation is, indeed, but a very remote cousin of the sound played. Countless details, both consciously applied and uncontrollable, enrich the note's sound and transport traces of the musician's education, the instrument used, and the contingencies of the moment of performance. Direct creation of audio, derived from real recordings, is therefore a promising way to overcome what is often perceived as a "machine-like" quality of computer-generated music. Doing so, however, is currently still tied to resources available only to a small group of researchers who understand the exceedingly technical vocabulary, and have access to costly hardware needed for their experiments. As the authors of the study note, even with their top equipment they needed nine hours to generate one minute of music, and they had to train their model for *weeks* using hundreds of dedicated computational units each of which costs thousands of dollars.

Such computational resources are, of course, not currently available to the ordinary consumer whose aim is, for example, to have their smartphone simply generate some soothing background music for relaxation. And even big companies have to be careful with the amount of computation that their machine learning projects generate on their servers (and, ideally, they also should care about the environmental impact of these computational projects). The next chapter offers several examples of the use of machine learning and other mathematical methods in relation to music, taken from actual everyday cases and grounded in the reality of our current technological situation.

Notes

1 John D. Kelleher, *Deep Learning*, 138–140.
2 It is true that some musical practices function without the need, for the performer, to have explicit musical-theoretical knowledge. Examples of such music-making range from football chants to lullabies. Yet, it is questionable whether these musical practices would have existed in their current form without the centuries-long influence of explicit musical theory, incorporated in

the audible structure of common music. Also, theory-less music is relatively simple, and, in the case of football chants, mostly directly derived from well-known melodies.

3   Ian Simon et al., "Generating Piano Music with Transformer," *Magenta*, September 16, 2019, https://magenta.tensorflow.org/piano-transformer

4   Cheng-Zhi Anna Huang et al., "AI Song Contest: Human-AI Co-Creation in Songwriting," *Magenta*, October 13, 2020, https://magenta.tensorflow.org/aisongcontest

5   John D. Kelleher, *Deep Learning*, 245, 246.

6   Some promising results have been achieved recently with regard to systems that generate visual output.

7   https://openai.com/blog/jukebox/; Prafulla Dhariwal et al., "Jukebox: A Generative Model for Music," *ArXiv:2005.00341 [Cs, Eess, Stat]*, April 30, 2020, http://arxiv.org/abs/2005.00341

11
REAL-WORLD MUSIC AI

It's an understatement to say that the world of technology companies is fast-moving. By the time this book appears in print, the businesses I discuss here will almost surely have morphed into something at least slightly different. A small company can potentially grow into a global success, changing its direction and product along the way, and a big company can, theoretically, disappear overnight. Yet, the technical trends that the examples in this chapter represent and the musical-mathematical approaches they use will still apply even if the fate of these businesses changes for the better or otherwise.

The first real-world case of mathematical music I would like to discuss here comes from the city I live in, Berlin. Here, a group of young businesspeople created in 2018 a company whose most important product at the moment is an application that generates background music.

The young founders who speak to the press about their company's product avoid calling the sounds generated by their system "music." They insist that it is not music but something that is not supposed to be listened to actively (as if this were the criterion by which people judge if something is music or not). Instead, they claim that the acoustic product they sell is "functional" in the sense that it enhances productivity or sleep through what they see as a scientific application of sounds.[1]

This company's case is interesting on several grounds. First, the founders know about the history of process- and randomness-oriented approaches to music from the 20th century, and they reference them in their public

DOI: 10.4324/9781003229254-14

representations of their product. Thus, pop-avant-garde composer Brian Eno's ideas about creating not the music itself, but a system that generates it crop up in their interviews, alongside post–World War II aleatoric music such as Terry Riley's piece called *In C* in which musicians in the ensemble simultaneously make individual decisions regarding which snippets of music to play, and how often to repeat them.

Both ideas translate perfectly to programming: A music-generating software is indeed a system that gives rise to the actual acoustic output while influencing its character through some internal qualities of the program itself. At the same time, random or probability-based playback of predefined musical snippets has been the basis of music-generating systems since at least Athanasius Kircher's 1650 wooden box with notation written on slats, and it has remained an indispensable component in such experiments throughout the centuries, from the *musical dice game* attributed to Mozart to, indeed, Terry Riley's *In C* and beyond (see the chapters on the early modern period and the 20th century for more context on historical parallels).

Another technique from 20th-century academic music that was mentioned by one the founders in an interview is the combination of instruments or voices performing the same material, but at slightly different tempos. This leads to a repeating cycle of first gradually growing distance between the sounds, and then of gradual rapprochement. This technique, called "phasing," is, together with the overlay of repeating snippets of different length, one of the two staples of 20th-century musical experiments, which range from Joseph Schillinger's theory of overlaying rhythms to Brian Eno's famous piece *Ambient 1: Music for Airports* which he created with tape loops of different length.

The machine learning, or deep learning, or AI, component is the newest weapon in the arsenal of the company's music technology team. It is difficult to be technologically precise here, as documentation of the program, let alone its source code, is, as often in such cases, not publicly available. In interviews, the founders mostly speak of AI in general or, sometimes, more specifically of neural networks. An example one of them has given is that he provided a network with hundred melodies, and received hundred new melodies in response, of which he discarded fifty as "unmusical."

It is not easy to say whether this description is so imprecise (just look at it in the context of the previous chapters on deep learning) because of the need to communicate with journalists in strictly non-technical terms or because machine learning is in fact not the central technology for what companies like this one try to achieve, but the latter option is theoretically possible, at least at this specific moment in this company's trajectory. While the attention-activating buzzword of "AI" is a good way to create interest in potential users and investors, deep learning might in fact be completely unnecessary to create a non-repeating soundscape whose musical param-eters are influenced by data about the listener's location, time of day, or pulse, like the background music generated by this program (but one *can*, of course, use machine learning even when one does not really have to, hoping that the extra effort would lead to fancier results).

As we have seen in the previous chapter, machine learning is good at in-ferring hidden features in the input data that lead to the desired output, as well as segmenting large amounts of data according to similarity, and at op-timizing the behavior of an artificial "agent" in a controlled environment. With enough ingenuity all of this can be used to create an ever-changing soundscape, but good old-fashioned randomization might also do the trick in many situations, and it can even be more reliable in a real-world scenario where consumer devices have limited calculating power and battery life.

One last striking characteristic of the discourse in which the company's founders present their product deserves mentioning here before we move on to the next example: their use of scientific imagery.

"Science" is, as of this writing, one of the top menu items on the com-pany's website.[2] The short presentation given there envelops the reader in real, but mysteriously-sounding language of professional musicology and biology like the "pentatonic scale" (a musical scale with five notes), "pure intonation" (a musical tuning based on ratios with alleged special aesthetic qualities—see the chapter on antiquity), and "circadian rhythms" (the bi-ological 24-hour cycle of sleep and wake time). Elsewhere, founders speak of "ultradian rhythm" and "sleep onset period," stress how their product employs "the latest research on how specific frequencies, musical scales, and phrases affect the person's cognitive state," and point to "scientific principles and research incorporated into the platform."

Against this background of references to modern scientific language, it is interesting to see how the centuries-old concept of the calculating mind continues to crop up in interviews of the company's founders. Describing their early experiments, one of them says: "We started with simple ratios of two tonal frequencies like octaves, 2:1, or a perfect fifth, 3:2, because those are pleasing to the brain. A new model suggests music is found to be pleasing when it triggers a rhythmically consistent pattern in certain auditory neurons." This is exactly the idea of the mind being more happy with calculating more simple relationships in its pursuit to create a model of the outside world that has been wandering through European aesthetics for ages,[3] but this time it is paired with the language of biology (for more historical background, see the description of contributions by Euler and Leibniz in the chapter on the early modern period).

Such framing of musical experience in timeless, culturally unaffected, and objective terms also reminds of the discourse of the Russian aesthetics research of the 1920s, as well as of Joseph Schillinger, the Russian-American music theorist who claimed, in the 1930s–1940s, to have found the "scientific" basis for music (more information on Schillinger is contained in the chapter on developments since 1950). Like their predecessors almost hundred years ago, this company's founders, who also come from Russia, are determined to marry aesthetics, physiology, psychology, and acoustics in their pursuit of a "functional" music machine. Will they succeed in curing the illnesses of the modern world like stress and difficulties in sleeping by using even more technology in these areas?

Now, let's turn our attention to a company of a, currently, completely different size. Having started years ago as a relatively small business, it is now difficult to overlook. With hundreds of millions of users, and with a revenue of several billions, its audio streaming service is well-known around the world.

Whereas the previous company endeavors to generate new, live, unique, and individually tailored soundscapes, the streaming company's business seems to currently mostly consist in providing already existing music and podcasts. Yet, the way the company structures the experience for the user is where the AI comes in.

The listeners can use the platform to find, by searching for a specific group, song, or podcast, an audio recording they already know about. They can also organize recordings in playlists and share these playlists with others. All of this can, in principle, be provided without using a machine learning component even if employing it might render some of the functions (such as searching with incorrect spellings) more comfortable.

Yet, having spent a lot of money on licensing all kinds of music and spoken content, the streaming company also has an interest in presenting its catalogue to the user. It is important here for the company to try to make the user see the recommendation as an added, valuable service rather than an annoying ad. The recordings offered to the user for consideration need to reflect his or her genre preferences, but also such subtle effects as long-term listening behaviors where listeners, sometimes, focus heavily on one portion of the music landscape just to move on abruptly to a completely different one.

This task calls for a machine learning approach. On the one hand, predicting the likeliness that a specific user who has listened to specific recordings in the past might also like recording X looks very much like the basic machine learning exercise of providing input data (different recordings) and corresponding output data (how much the user played them) to let the system infer the hidden relationship between the two.

However, exactly what kind of input data one can use is also important. Obviously, the database of the service contains band names and titles of songs and albums. But is it already enough to reliably recommend further titles? Since we already know that machine learning systems tend to give more accurate results with more data, we might want to look for additional parameters.

A presentation that was given by two of the company's machine learning researchers in 2017 sheds some light on what goes on under the hood of its recommendation engines at this stage in the user's interaction with the system.[4] On the one hand, it is possible to rely on additional manually created data about the recordings that is stored in the system, including genre labels. On the other hand, it is now also possible to automatically extract different characteristics of the recording from the audio itself, and to use them as data in the machine learning recommendation system.

A programming interface which has been made available to the general public by the company offers a glimpse of the power of automatic feature extraction to surround a musical recording with a plethora of numerical values and labels which are all potentially valuable in a machine learning environment.[5] For example, it is possible to extract, for more or less common music written and performed in a Western idiom, not just the tempo (in beats per minute), but also how likely it is that a piece is performed with acoustic instruments, whether it contains vocals, whether a live audience is present in the recording, and even if a track sounds "positive" or "danceable," in addition to features coming from Western music theory such as key or time signature.

It is clear that dealing with these mountains of data produced by the system and by the users' interactions with it invites machine learning approaches, including not only the kind of supervised learning outlined above, but also unsupervised clustering of users and tracks into groups according to similarity and, possibly, even reinforcement learning aimed at maximizing the goals the company has in its interactions with the listeners (see the previous chapter on the differences between these types of machine learning). For recommending spoken word content like podcasts, automatic speech transcription enabled by machine learning absolutely suggests itself, and if it is not already being used today, it will surely join the company's army of algorithmic tools in the very near future.

When a group of academics interested in studying the technological and social side of the company's business model did research on its job listings, they found that in late 2016 and early 2017 alone a whole 34 out of around 500 new jobs were centered on big data and machine learning.[6] These jobs were scattered throughout many of the big company's departments, including advertisements, but it was clear that a portion of them would be aimed at employing machine learning for recommendations.

A talk given by an employee in 2019, just two years after the presentation mentioned above, shows that the company's machine learning component has since indeed grown into an engineering effort of industrial scale.[7] No longer able to tolerate the potential destabilization of its computer system by individually crafted programming experiments, the company has moved to a standardized machine learning infrastructure. The goal of this transition is to help the engineers who oversee the gigantic, constantly updating and changing system to train the networks faster and to catch errors more easily.

The two different courses taken by the companies that I used as examples in this chapter make evident the seemingly obvious difference, from the point of view of the listener, between generation and recommendation. In one case, new audio is created by the system whereas in the other case existing audio is recommended.

However, these two areas do not have to be separate. It is entirely thinkable that, given a large enough pool of recordings, the boundaries between them could become blurry. A system aimed at providing relaxing or activating audio on the basis of some external data such as time of day could, in principle, choose one out of thousands of existing recordings instead of generating new audio. Likewise, the music recommendation engine that has taken in the user's listening trajectory and tastes can, in principle, provide parameters for generating new pieces instead of choosing existing tracks.

To conclude this chapter, I would like to point out another area of musical use of machine learning which lies somewhat outside of both generation and recommendation. In a professional music studio environment, a recording goes through several stages, each of which needs attention from a specialist. Recording engineers, for example, employ a lot of knowledge and experience to choose the right kind of microphone, to position it correctly, to oversee the loudness and frequency range of the recorded voice or instrument, and to do many other kinds of studio work. Another tricky procedure is mixing the individual voices and instruments together, while at the same time adjusting sound parameters and effects such as reverb, or echo.

But the holy grail of studio work, accessible only to the most elite specialists with the most expensive studio rooms and hardware is *mastering*. Here, the already mixed recording is treated by studio effects once more, ensuring that it will sound good on different kinds of consumer devices, and that its loudness and character fit the expectations of its genre.

Given this initial situation, to many it came as a kind of a shock that automated machine learning solutions for mastering began to crop up recently, causing a real stir in the studio recording community.

One of the examples of such attempts at automated mastering comes from a company that has since put much effort into becoming a more broadly defined platform for musicians, including the automatic publishing of tracks on streaming services.[8]

Academic research on this company has been so far characterized by the same difficulty that I have mentioned above of stating exactly which kind of machine learning is used by which part of a company's technical set-up.[9] In future, it seems possible to learn more about it as both the company might grow, with its employees giving more technical presentations at specialist conferences, as in the case of the streaming company, and as more work on the possible origins of its techniques in previous academic research is carried out by historians of technology.

Meanwhile, it makes sense to keep in mind the idea that changing technology might not in fact be the most important driving force behind companies that use, or claim to use, machine learning as a tool in relation to music and sound. Instead, the way the companies present themselves, their course of development, and even their choice to use machine learning at all rather than some of its less trendy alternatives, are arguably most deeply influenced by the economic and cultural situation in which the companies operate, with the need to cater to the interests of venture capital investors as possibly the single most important factor.

In an attempt to make their product seem like a viable replacement for an existing, but more expensive, practice, companies in fact try to redefine the very kind of activity they pretend to merely replicate.[10] Ironically, this might be exact equivalent of what was happening with mathematics and music for centuries anyway (see the chapter on putting music AI in perspective).

Notes

1 For quotes discussed in this and in the following paragraphs, see the interviews by two of Endel's founders, Oleg Stavitsky and Dmitry Evgrafov: Tyler Hayes, "The Science behind Endel's AI-Powered Soundscapes," *Amazon Science*, November 25, 2020. https://www.amazon.science/latest-news/the-science-behind-endels-ai-powered-soundscapes; Denis Bojarinov, "Dmitry Evgrafov, "Iz sta melodij, kotorye mne sdelala nejronnaja set', ja polovinu vykinul," *Colta. ru*, August 5, 2020. https://www.colta.ru/articles/music_modern/25073-dmitriy-evgrafov-intervyu-albom-surrender; Anastasiya Skrynnikova, "'My ne sozdajom muzyki': osnovatel' servisa generacii zvukovogo fona Endel o sdelke s Warner Brothers i rabote algoritma," *vc.ru*, April 10, 2019. https:// vc.ru/future/63710-my-ne-sozdaem-muzyku-osnovatel-servisa-generacii-zvukovogo-fona-endel-o-sdelke-s-warner-music-i-rabote-algoritma.

2 https://endel.io/

3 Robert Wicks, *European Aesthetics. A Critical Introduction from Kant to Derrida* (London: Oneworld, 2013).

4 Ching-Wei Chen and Murali Vidhya, "Machine Learning and Big Data for Music Discovery at Spotify. Presentation." Galvanize NYC. March 9, 2017. https://www.slideshare.net/cweichen/machine-learning-and-big-data-for-music-discovery-at-spotify?qid=b7ca7727-0a01-4441-9107-14410ccf0e7d

5 "Get Audio Features for a Track," Spotify for Developers. https://developer. spotify.com/documentation/web-api/reference/tracks/get-audio-features/

6 See the chapter *Intervention: Work at Spotify!*, in: Maria Eriksson et al., *Spotify Teardown: Inside the Black Box of Streaming Music* (Cambridge, MA: The MIT Press, 2018).

7 "For Your Ears Only: Personalizing Spotify Home with Machine Learning," *Spotify Engineering*, January 16, 2020. https://engineering.atspotify.com/2020/01/16/ for-your-ears-only-personalizing-spotify-home-with-machine-learning/

8 https://www.landr.com

9 Jonathan Sterne and Elena Razlogova, "Machine Learning in Context, or Learning from LANDR: Artificial Intelligence and the Platformization of Music Mastering," *Social Media + Society* April–June 2019: 1–18. https://doi. org/10.1177/2056305119847525; Thomas Birtchnell, "Listening without Ears: Artificial Intelligence in Audio Mastering," *Big Data & Society* 5, no. 2 (July 1, 2018): 1–16. https://doi.org/10.1177/2053951718808553

10 For a historical account of how such redefinition of human activity into a mechanizable procedure already took place during the early decades of AI research, see: Stephanie Dick, "Of Models and Machines. Implementing Bounded Rationality," *Isis* 106, no. 3 (September 2015): 623–634.

12
MASS-PRODUCED AND STILL INDIVIDUAL

For a long time, mass culture basically meant the sale of a large number of identical copies of the same cultural product. Aided by AI, popular music might become the opposite of that: The production of a large number of cultural products that are all *different*, because each of them is optimized for a single listener.

Mass-produced and individualized. Aren't these words incompatible? Not anymore, at least not with regard to digital products. Thinking about mass production, one is tempted to imagine the factories of the industrial era that turned out, year by year, the same physical product because it was prohibitively expensive to reconfigure them. But the digital tools of production are much more malleable, so much so that it is often cheap enough to use them for one-off tasks. Examples of such individually optimized products or services are in fact so numerous that it is increasingly becoming difficult to find a place on the internet where they are *not* used.

Social networks and search engines gather information about their users and present, based on this information, highly personalized reading feeds and search results. Online shops do the same to recommend products that they hope the visitor would buy. Automatic filtering out of unwanted spam messages has become an almost universal feature of email systems, and some providers even go as far as to try to find, for the user, among the *non-spam* messages the ones that are most important.

Businesses built around existing music already use a data-based approach augmented by machine learning (see the chapter on real-world music AI).

DOI: 10.4324/9781003229254-15

Experiments in computerized music generation have been around since decades, and their results become more and more convincing. Whether the music business *as a whole* will now combine these two trends and turn towards automated generation of content depends on how well the public will react to it. But someone is surely going to try.

During the previous decades, a commercial music producer semi-intuitively tried to orient their product towards an imagined mass audience or an imagined segment of that audience. A certain amount of generalized statistics was available such as the information contained in the charts about the sale and airplay of records. But this information was about a large group of people, so it did not tell the producer anything about the personal preferences of a specific listener. Now, things have changed dramatically, at least concerning the collection of data. A music streaming service now has an incredibly detailed picture of the listening interests and habits of each and every of its users. This broadly applied commercial surveillance could form the basis for made-to-measure music whose attractiveness for the only person who listens to it is almost ensured by everything the system knows about the person.

The "only" problem (provided that one ignores the cultural or ethical implications of such a situation) would be that somebody would still need to create the music, based on the parameters calculated by the system, and this labor could become more expensive than the revenue generated by the act of listening to the result. But then again, with all the computational composition tools, this work could be automated to such a degree that creating a music recording would become not more expensive than creating and sending a spam message. Add to that the possibility of rich producers exploiting the work of vulnerable digital workers around the world, and we end up with a pretty grim picture of a future music business.

On the other hand, it is well known that attempts at predicting the future are notoriously error-prone. They are indispensable as a tool to think through what we want our future to be like, and which scenarios we would like to avoid, but it would be illogical to assume automatically that the worst possible outcome is also the most probable one.

For the time being, it is safe to say that computational tools will enter the commercial workflow in music—but only as a part of the bigger picture. It is unlikely that the music business will be wholly automated in any foreseeable future. Data-based, highly individualized, yet still only semi-automatically generated products may, however, proliferate in the world of entertainment.

Academic debates around creative uses of computational technology have been going on since decades. There, the idea that an augmenting, or assisting, role of the technology is more probable than an outright replacement of the human actor is widespread. So far, the reasons for this belief were mostly technical—the poor performance of the systems on creative tasks without clearly defined inputs and goals.[1] But, as systems start to perform better, the reasons for not using them might become cultural.

Whether any of what I try to imagine here is actually going to happen largely depends on future shifts of cultural factors. One of the biggest obstacles in the way of those hoping to use automated composition in a mass-cultural context might be the current practice of shared identification with the performer. Identities constructed around shared consumption choices, or, to put it differently, the identification with the social group that makes the same choices (e.g., listening to a specific star performer) seem to be antithetical to the practice of individualized entertainment. In the world of personalized music there would also be no income from concerts (as they would become impossible due to the lack of a shared musical product), and maybe even no such thing as a star interview, as a stable (performed) identity of a music star would become much more difficult to uphold in a situation where each recorded piece is different and meant for only one listener.

The future also, and crucially, depends on the public's general attitude towards automatically created cultural products. The musicians of tomorrow will have to relate to what listeners imagine to be an adequate, and acceptable, music-making machine. Today, listeners generally do not see a musical clock as a legitimate instrument that allows the musician to express their creativity. By contrast, an electric guitar connected to a set of sound effects and a sophisticated amplification system is universally accepted as a musical instrument, despite the fact that the amount of "work" done by technology is also high in this case. All these perceptions might shift in some new ways in the future, and they will probably have to do so to accommodate the new semi-automated world of entertainment.

The public's ideas about musical machines also reflect their long history. As I have shown in the first part of this book, music-generating mechanisms were first imagined in a pre-digital, mechanical world. Going back centuries, they even existed before the sound-reproducing technologies such as the Edison phonograph became available. Any attempt to ignore this cultural heritage and to force the listener into a completely antithetical new development is likely to fail. Instead, an attitude of "everything old is new again" looks like a much more viable plan.

So why do I think that creators and listeners of popular and commercial music will generally accept tools grounded in data and AI? Because there is an existing discourse that legitimizes it and assigns value to the products of such a procedure.

This is the discourse that has surrounded rule-based composition and music games for centuries, the belief that such an approach to music is both possible and helpful. Basically, it's the idea that music is mathematically structured sound. As I have mentioned at the beginning of this book, it is possible not to subscribe to this notion. But today it is as powerful as ever, and a big part of the creators and listeners seem to believe in some version of it.

Now, what are some of the possible near-term future developments? It is quite possible that the musical use of machine learning and data will enter the listening public's horizon in the same way as other technologies of the past: through unforeseen and unintended use, and as a novelty. This has happened with turntables when they were repurposed for the creation of beats and for scratching, and this has happened with Auto-Tune, a technology initially meant for correcting singing intonation errors, and famously later instead employed as an effect.[2]

Also, without even going into the territory of personalized recordings, one might think of a tool for creating prototypes of commercial music based on predictions about the future changes of the music market. It takes some time to record (especially, when human performers are involved) and to make a piece of music well known to a large enough chunk of the audience. Making a decision to record a specific piece today, the producer actually creates something for the listener of tomorrow.

In this context, it is possible that more and more producers and music labels would try to join the bandwagon of data-based predictions: Looking at the dynamics of music trends, they would attempt to predict which trend in music is worth investing into given the time needed for production and everything else. And while one can try to achieve this goal of fortune-telling using only simple statistics and information visualization (such as the typical line chart) there is no rule that forbids at least trying to use more sophisticated tools, possibly the ones used in weather forecasting, or even in automated trading. Several companies already strive to offer services of collecting and analyzing music market information with the goal of facilitating business decisions such as finding promising artists.[3]

The deadly coronavirus pandemic that has recently reshaped so many facets of everyday life around the world will surely also contribute to an acceleration of this shift. Reliance on predictions and modelling based on statistical data has now become such a natural part of day-to-day functioning of billions of people that it is unimaginable that this will not make data-driven music business seem like a more acceptable idea. Also, as mass concerts have remained impossible during the pandemic, and music streaming has replaced some of this cultural practice, the shift to individualized musical entertainment has begun to seem more and more imaginable.

As mentioned above, it is unlikely that the whole "pipeline" of commercial music would become automated. Instead, it is, for example, conceivable that a producer could use the prediction system to set the parameters for generating a couple of prototypes, from which they would then choose the one to be produced more professionally. In doing so, they could choose to select the features more intuitively, by looking at data on successful recordings of the chosen type of music and on the potential listeners, or they could delegate this choice to a machine learning algorithm, hoping that it would successfully optimize the musical product's features to the desired audience.

And if this sounds like some kind of musical dystopia to you, you are probably not alone. Nobody wants to feel spied on, stereotyped, and manipulated into buying a product. One does not need a crystal ball to predict that this kind of over-optimized synthetic music will promptly lead to a cultural backlash and a counter-trend of handmade anticommercial music, which will then in turn become co-opted and imitated by commercial producers, and so on, and so on, in the same way as it has been happening in popular music for decades. It is even thinkable that producers who use data and machine learning could purposefully underoptimize their product, adding slightly unfitting elements, in order to avoid creating in the listener the negative feeling mentioned here.

A data-based approach such as the one described above would not be a completely new thing. In fact, the world of entertainment already made some significant steps in this direction in the past, preparing the ground for the full-scale future adoption of data-based decision making. There is the whole decades-old industry of adjusting television content to ratings. There is an even older tradition of test screenings in Hollywood. So-called focus groups (group interviews) are a well-known instrument of market research. Data from social media are routinely gathered and analyzed by different actors in the music market, and, before the coronavirus pandemic broke out in 2020, it was seemingly permissible to refer to advertising campaigns involving actions by users of social media as "viral marketing."

Even without the use of data and mathematical tools to choose the features of the future musical product, the world of popular music can sometimes feel very industrial, and, in fact, machine-like. Nowadays, groups of musicians, each specializing in one specific facet of contemporary pop writing, often convene to contribute collectively to a musical product. According to one study, the hits of 2016 were written by 4.53 musicians *on average*, with the more extreme cases crediting up to twelve persons.[4] In such a scenario, the variety of musical genres mixed together into one package connects to the tastes of a broad audience, as well as to the many different musical interests each individual listener now has due to the constant availability of even the most niche music on the internet.

In 1977, the film *Star Wars* famously combined very different film genres into one irresistible mix. One of the more impressive early examples of a musical cocktail of comparable heterogeneity is the song *We've Got it Goin' On* by the boy group *Backstreet Boys* from 1996. It was created by a whole team of writers, and has been analyzed as containing in itself traces of ABBA-like melodism, 1980s arena rock choruses, and 1990s R'n'B grooves, as well as nods to both the 1986 hit *The Final Countdown* (the synth horn sound) and the 1875 orchestral piece *In the Hall of the Mountain King* by Edvard Grieg (the staircase-like main melody)[5]—and I would also add Michael-Jackson-like singing style to this list (the rap part). No one person is capable of contributing all of these aspects. The division of creative labor that stands behind hits like *We've Got it Goin' On* is only one step removed from becoming a semi-automated workflow, and this final step would be to replace a human coauthor with a program.

One of the more famous early instances of data-based business decisions in the world of entertainment was the launch of the successful series *House of Cards* in 2013. Here, again, several disparate aspects—a popular director, a well-known actor, and a well-received previous series—were mixed together to form what proved to be a hit. Yet, in this case the decision was (allegedly) made based on the company's detailed knowledge of consumer behavior.[6] And given that the company is involved in not just production, but also, and primarily, streaming, this knowledge might have been not only about a statistical field, but also about individual viewers. A single product was then still made for a mass audience. But it is not inconceivable that at least slightly different versions of a program could be made for individual viewing in the future, based on everything the producers now know about us.

Let me now elaborate a little more on the exact reasons why I think that a semi-automated world of entertainment is not far away in the future, but a fully automated music business "pipeline" is unlikely.

First, the world of commercial music consists not just of products (recordings), but also, and primarily, of processes that lead to the creation and dissemination of these products. Behind every song, there are people who make myriads of choices, including the choice to adopt or to ignore a certain technology. People working for music labels, producers, performers, as well as people involved in concerts, sales, or advertising, will resist the disruption of their work and livelihood. Their voices of protest might reach the listeners, and might lessen their willingness to adapt to the new, semi-automated paradigm. Without the cooperation of today's creators, the new technologies could be perceived by the audience as a threat and not as a potentially interesting and helpful novelty.

You, the reader, might have already started to wonder why my description of the future leans so heavily towards the language of business and entertainment. Aren't purely artistic goals, beautifully difficult works of art, or conceptual playgrounds interesting? Of course, they are. But my main reason for concentrating here on the world of money and entertainment is... technical.

As mentioned above, solutions based on machine learning thrive especially in situations where there is a clear measure of success. And, in a commercial project, it is much more clear whether something was a success or not, and if yes, to which degree.

A whole world of very valuable non-commercial artistic experimentation exists in which mathematical and technological tools are used with a great deal of ingenuity and cultural purpose.[7] There, one can set specific optimization goals (such as, for example, imitating a certain historical style), and then do conceptual and artistic work with a system created in this way.

Yet, the decision whether such a project was successful remains very subjective. A project might even fail completely on the technical level, and still be seen as an artistic success precisely because of its shortcomings that now can be used for a specific effect.

By contrast, design decisions become much easier if one turns towards the more clear-cut ways in which commercial projects operate. By looking at the music business, one avoids the technical problem of not having an obvious numerical measure for success. Here, the number of plays and other metrics such as mentions on social media and inclusion in streaming playlists can be used as signposts. This simple observation, of course, does not mean that music created outside of the world of entertainment cannot be created through machine learning. But it means that the limitations of the existing technology privilege some uses over others.

In the boardgame Go, the number of points each player has at the end of the game is a clear numerical indicator of success. With it, the program knows exactly whether the new version of its network settings delivers a better, or, conversely, worse result than the previous one. Go was one of the early public breakthrough moments for machine learning. Commercial music might become the next one.

From all the possible future developments one seems especially interesting for me. It is an approach for which I do not have a better name than "harvesting." The core idea is thus: A producer generates a great number of different versions of a musical product, almost at random, and launches them all at once. Most of them will inevitably remain ignored, but the hope will be that at least one or two will start to grow in popularity. At this point the producer will concentrate on these candidates for further development, and start to invest more resources in them. Like a start-up or an early-career pop artist, each variation will receive a brief chance to try things out in the wild. At its core, this is not a new scheme. It has been employed across the spectrum of human creative work for a long time. Websites, for example, often try out new design features on a section of the audience, and compare the results with the reaction of the users who see a different version.

But the crucial new element here is that it does not have to happen in real time and with real listeners. With all the data about listeners' habits and interests, it is also thinkable that the generated music might be tested not against the real market, but against a machine-learning-created model of it. Then, only the survivors of this test run would be used in production.

Decades ago, programmers became enchanted by the idea that one could create artificial life, happening entirely within a program and its data. Experiments ensued, with numerous imitations of the natural process of random mutation and the survival of the fittest. There were even games built around this idea.[8] Soon, producers might begin to play games with music, and try to offer you, the listener, the result of their "breeding" and "harvesting" endeavors.

Another possible future experiment might be the creation of individually generated celebrity products. As mentioned above, the stability of the performed identity of a music star is endangered by the practice of individually created recordings because each listener basically has their own version of the star and the music. But music stars from previous eras who gained popularity during the time of mass-produced, non-individual popular music could try to benefit from the new paradigm. It is therefore thinkable that an existing celebrity might try to sell a new "song" in which unique variations are created for each purchaser, based on data on their music tastes, demographics, or even search history or other data amassed through commercial online surveillance. Whether such an experiment might actually harm the image of the celebrity remains to be seen.

In the end, technical and cultural change will likely lead to the emergence of a new musical aesthetics—one that fits the new technology of machine learning.

When, in the first half of the 20th century, new production technologies became widely available for the manufacturing of furniture and housewares, a new, modernist aesthetics of everyday objects followed suit. Objects with geometrically simple forms and clear colors became fashionable not just because they differed from previous styles, but also because they fitted the new, industrial era of production. Basically, you could mass-produce a very nice spherical teapot, but you could only mass-produce a very poor and crude copy of a Baroque bowl. The German arts and architecture academy Bauhaus was among the defining institutions where this change took place. The Soviet academy Vkhutemas was another. But soon, the new aesthetics took over the whole world, at least until it, too, came to be seen as a temporary wave.

The same could happen with AI and music. Production tools might become available that allow the creation of beautifully crafted pieces—but only if they fit a certain aesthetics, which might be very different from what we today expect from popular music.

During the 1920s, Bauhaus and Vkhutemas were important institutions, but they were neither centered on music, nor were they the only ones trying to tease out a new, modern aesthetics. Especially during the early decades of the 20th century there were dedicated music research institutes that looked for a new, scientifically grounded, musical practice. The next, final chapter of this book is dedicated to their lasting impact and the connections that exist between their discourse and the language of today's music AI.

Notes

1 Artemi-Maria Gioti, "From Artificial to Extended Intelligence in Music Composition," *Organised Sound* 25, no. 1 (2020): 30. https://doi.org/10.1017/S1355771819000438
2 Melissa Avdeeff, "Artificial Intelligence & Popular Music: SKYGGE, Flow Machines, and the Audio Uncanny Valley," *Arts* 8, no. 4, (2019): 130. https://doi.org/10.3390/arts8040130
3 Rocchi Tommaso, "How Data Is Redefining the Role of A&R in the Music Industry Today," *blog.chartmetric.com*, September 21, 2020. https://blog.chart-metric.com/what-is-a-and-r-music-industry-today/

4 Mark Sutherland, "Songwriting: Why It Takes More than Two to Make a Hit Nowadays," *Music Week*, May 16, 2017. https://www.musicweek.com/publishing/read/songwriting-why-it-takes-more-than-two-to-make-a-hit-nowadays/068478

5 John Seabrook, *The Song Machine: Inside the Hit Factory*, 2016, 75.

6 David Carr, "Giving Viewers What They Want," *The New York Times*, February 24, 2013. https://www.nytimes.com/2013/02/25/business/media/for-house-of-cards-using-big-data-to-guarantee-its-popularity.html

7 The field of high-cultural artistic use of programming and, specifically, neural networks is huge and extremely diverse. For an overview, see: Alex McLean and Roger T. Dean, eds., *The Oxford Handbook of Algorithmic Music* (Oxford: Oxford University Press, 2018).

8 Some of the games built around this principle are *El-Fish* (Maxis 1993), *Creatures* (Mindscape, 1996) and *Spore* (Maxis, 2008).

13
AVANT-GARDE BECOMES POP'S AIDE

In 2018, one of the co-founders of a music generation start-up presented his company's system to the public in a short, entertaining talk.[1] By that moment I have already begun to research musical uses of machine learning and the culture that surrounds it, so I was prepared to hear a presentation that paints the technology in somewhat romanticized colors and gives it more magically-sounding agency than what it actually deserves. But the talk surprised me in a way I could not predict. Some of the language used to discuss the system seemed to resemble the discourse of early Soviet modernist aesthetics research, from one hundred years ago.

The machine learning system was presented in the talk as an "artificial intelligence that has learned the art of music composition by reading over 30,000 scores of history's greatest," a program that has automatically derived its mathematical rules from the "scores written by the likes of Mozart and Beethoven." This has allegedly enabled the system to compress the process of learning, which takes humans years and decades of work, "down to a couple of hours."

I was intrigued by the similarity I saw between this picture of musical AI as grounded in a corpus of existing works, as well as in statistics, and mathematical science, and the ways in which early Soviet musicologists tried to establish new music and new musical tools in the 1920s.

As mentioned in the first part of this book, some of them tried to work in a similar vein, hoping to derive universal musical "laws" from an analysis of a very large number of works. Konjus, for example, looked at almost

DOI: 10.4324/9781003229254-16

1,000 works, including all Beethoven symphonies and sonatas, in his hope of showing that all good musical form is based on simple proportions. Sabaneev seemingly analyzed almost 2,000 works in his attempt to show that good music is built around the Golden Ratio proportion.[2] Schillinger also claimed to have inferred his mathematical composition rules from an inspection of the masterpieces, thus enabling the users of his system to work much more quickly and efficiently (see the chapter on developments since 1900).

Without the tool of the modern computer, and without all the digitally encoded scores which are now plentiful on the internet, the Soviet researchers had to do everything by hand. As can be seen in the examples provided above, some of them also had pretty exotic and narrow musical theories which they simply tried to "prove" by finding fitting examples from music history (in a machine learning situation, this would have been called a technological bias—a situation where the system is too simple or inflexible to really adapt to the training data).

Still, the general approach—taking thousands of works of "great" composers, turning them into a mathematically derived representation, and using this representation to generate new music—was the same as it is now with music AI. The basic idea that supports such thinking is that musical intuition (something that past composers used to create their masterpieces) is actually unconsciously grounded in mathematics.[3] Once one subscribes to this notion, it starts to make sense to try to look for hidden mathematical structures in famous pieces of music, and to try to imitate their success by producing more pieces with the same mathematics in them.

The work done by Konjus, Sabaneev, and their many similarly minded colleagues at early Soviet arts research institutes sounds decidedly modern, especially when one looks at its parallels not only to the language around musical AI, but also to the currently very active research area of the Digital Humanities. There, scientists use the statistical analysis of large collections of works in their hope of providing insights not possible with the more traditional methods of interpretation and of working with a single source.[4]

The work done in the 1920s also prefigured some of the post-World War II approaches that tried to unite aesthetics and information theory. Thus, Sabaneev in the 1920a attempted to calculate mathematically the "originality" of a musical work by measuring statistically the frequency with which individual motifs appeared in it, and Rozenov tried during the same period to popularize ideas about the need to have enough new information in a work to make it interesting, but not too much (overload), and not too little (boredom), coming remarkably close to the later ideas of the French theorist of computer art Abraham Moles.[5]

The reason why I discuss all these parallels between the past and now is that I find it really striking how so many contrasting, even contradicting things could become so closely entangled. First, there is the contrast between the heady and modernist aspirations of the people interested in mathematical music in the 1920s and the applied, practical goals of today's music AI business. The tools and the language of the avant-garde have become the working instruments of popular and functional music. Instead of casting away all traditions, the historical remainder of mathematical experiments of the modernists from 100 years ago has now become an aide of pop.

Second, I find it really funny to see (or to imagine, if you do not agree with my findings) the parallels between the language of early Soviet researchers, who were operating under strict political control from a communist dictatorship and of start-up companies, working under the restrictions of venture capitalism and short-term investments. The element connecting both worlds is the promise of efficiency, grounded in a "scientific" procedure and the "rules" derived from the works of the music history's heroes.

The early Soviet state had the hope of swift industrialization of its largely peasant population through the application of the most modern and efficient psychological and technical tricks. The creation of functional music, which was hoped to raise the spirits of the workers and to inspire them to greater achievements, itself also had to be streamlined and optimized. The promises of a novel scientific musical tool derived from tradition were too good to be ignored, and they helped the researchers find funding to pursue their interests even during the most difficult times of hunger and dictatorship, at least before Stalinist traditionalism really unfolded. Today, the promise of a machine learning system that derives all the important hidden rules of composition from a vast archive of the musical past, and does so more efficiently than a human worker, is the kind of idea likely to support a start-up company through the difficult early years of its existence. The same was true in the New York City of the 1930s where the poor, stateless Russian immigrant Joseph Schillinger had to create a new livelihood for himself by teaching American musicians the mathematical composition tricks that allegedly made their work so much more efficient.

If the music business is your way of survival, the promise of mathematical effectiveness is not to be ignored. In the 1930s and 1940s, as the American musical entertainment industry grew and became more professionalized, Schillinger offered exactly the kind of mathematical advice that people wanted to hear, despite the fact that he actually imported many of his ideas from his modernist Soviet past.

By doing so, he created a situation not normally seen in interactions between the avant-garde and the popular culture. It often happens that modernists distance themselves from the "primitive" popular culture. Sometimes, they play with it, ironically integrating bits of entertainment clichés and advertisement into their critically-minded works. But it does not happen very often that the popular culture integrates into itself pieces of the avant-garde, in the hope of becoming more efficient through the avant-garde's exotically mathematical methods.

The early Soviet research into mathematical aesthetics suffered an (often literally) deadly blow with the official switch, in the 1930s, to traditionalism and the so-called "socialist realism." Relevant work in this area stopped in the Soviet union for decades, and researchers previously interested in statistics or proportions now had to reorient themselves or risk oppression from the government. How, then, could the ideas survive and reappear decades later in a completely different environment?

I have three not mutually exclusive answers to this question: (a) direct acquisition of specialist knowledge, (b) independent rediscovery, and (c) intermediary discourses.

First, it is known that some of the people from Soviet arts research institutions did escape the Soviet Union and continued their work abroad. Sabaneev, for example, left the country in 1926, and Schillinger did not return from an official trip to the West in 1928. Schillinger also stayed true to his mathematical interests throughout his whole life and influenced a whole generation of musicians through his teaching in New York City and, later, posthumously through his theories which were taught during the early years of the now famous Berklee College of Music in Boston.

It is also possible that some of the early Soviet ideas, forgotten and marginalized because of the political tensions between countries at that time, were independently rediscovered in later years. The history of science and technology is full of such rediscoveries and parallel discoveries of little known approaches. And because the early Soviet mathematical aesthetics research relied heavily on 19th-century German work, it is very easy to imagine that someone could independently come to similar conclusions, just by following in the same path of German positivism.

But for me, one of the most mysterious, and therefore fascinating, ideas is that of a hidden intermediary discourse: someone secretly (or maybe unconsciously) borrowing from Soviet research.

Combining all three explanations, and without being able to prove it at this point (as this would have required writing a separate book), I would like to suggest the following speculative main route of influence (to which, of course, many additional routes must exist):

Avant-garde including Schillinger -> experimental electronic music including Stockhausen -> builders of popular electronic instruments -> users of these instruments -> a universally shared idea.

The first transition in this scheme is probably the most difficult to demonstrate. Luminaries of Western post-World War II experimental music like Karlheinz Stockhausen were reluctant to admit earlier influences,[6] and certainly especially so when these influences pointed towards the then current political enemy—the Soviet state. After all, their difficult and abstract music was seen as an antidote to both the aggressive traditionalism of Nazi Germany (the defeated enemy) and the dumbed-down triumphalism of "socialist realism" from the Soviet state (the upcoming enemy). But, as the readers of this book know, the Soviet culture was not always simplistic, and there are indeed certain parallels.

Stockhausen's article called... *How Time Passes...*, for example, is replete with diagrams and descriptions of various possible proportions in a work of experimental music,[7] which all call to mind the search, in the 1920s, for new musical tools grounded in mathematics. Among other ideas, Stockhausen describes there the combination of different meters like 9 against 7 or 7 against 5—the same idea that Schillinger used as an introductory exercise for all his American pupils in the 1930s and 1940s—and visualizes them in a diagram similar to a figure in one of Schillinger's publications.[8]

Also, the use of novel sound-generating technologies, did not first emerge in the 1950s, but goes back to widespread similar efforts in Russia, Germany, and elsewhere in the first half of the 20th century, and is predated by even earlier singular experiments at the end of the 19th century.[9] Yet, the circumstance that musical-mathematical experimentation took place in Russia and Germany during the rule of the respective dictatorships[10] must have made it impossible for Western post-World War II modernists to openly relate to these traditions.

The next transition is maybe somewhat easier to document. In a book-length history of the famous popular-music-oriented Moog synthesizer from the 1960s Stockhausen features a lot.[11] There, in quotes from the pioneers of American synthesizer culture his name often stands in for all "weird" experimental and otherworldly sounds that entered the public imagination at that time (following the short period of popularity of the Theremin and Trautonium electronic instruments in the 1930s).

At this point, it makes sense to briefly look at the technology of the synthesizer, and how it relates to the history of mathematical methods in music. The most famous kind of electronic sound production that became widely used in popular music is the so-called subtractive analog synthesis. At its core, a synthesizer using this technique contains a source of vibrations, and various means to influence the parameters of these vibrations, partially subtracting from their initial sound (therefore its name).

The whole mode of using such synthesizers is highly mathematical. To be sure, the builders of these machines did (mostly, and with some famous exceptions) include the traditional piano keyboard among the myriad controllers and switches with which the synthesizer could be played.[12] Without this nod to the tradition, it would have been difficult to sell the machine to a trained musician. But traditional keyboard layouts remained merely an input among many, and some of the other controls were actually better suited for a creative playing around with the electronic circuitry. It was these rotary controls and switches, labelled with numerical scales and, often, proportions, that arguably burned the idea of a mathematical music deep into the minds of millions of musicians and listeners of that "analog" era.

Figure 13.1 shows three rotary controls from the much larger control panel of the popular Minimoog synthesizer, first introduced in 1970. On the left, numbers from two to thirty-two denote the basic pitch region in which vibrations will be generated. Each consecutive number is twice the previous one, indicating the acoustic interval of the octave, defined by a ratio of 1:2. This way of describing the sound with a number is derived from an earlier musical technology, the organ, where the same notation consisting of a multiple of two and an apostrophe is used. The control in the middle of Figure 13.1 allows for a more fine-grained adjustment of frequency,

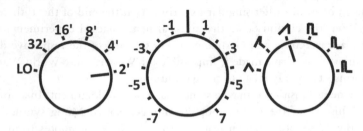

FIGURE 13.1 A diagram of a small part of the Minimoog analog synthesizer's control panel.

again expressed in numbers, and the control on the right changes the shape of the individual oscillations that make up the vibration (the so-called waveform). As can be seen in this example, the 20th-century musician was often required to think not only (and sometimes not even primarily) in terms of notes, chords, or melody, but of frequencies, numbers, and ratios.

Maybe the final step in bringing esoterically mathematical ideas about music into the minds of millions was achieved when *Switched-On Bach*, a recording of adaptations of Bach pieces made with a Moog synthesizer, became a hit album in 1969. There is certainly some truth to the widespread understanding of this album as simply a clever move by which the new (the synthesizer) was popularized and legitimized through the old (the canon of classical music). But it is also interesting to see that Bach, of all famous composers of the past, was used with great success in this endeavor.

Bach is often seen as one of the most mathematical composers. The features that contribute to this image of his music are, seemingly, (a) his famous use of a novel tuning scheme for a very popular cycle called the *Well-Tempered Clavier*, and (b) his mastery of Baroque counterpoint, a technique involving independent movement of several voices, all embraced by a common harmony and a whole arsenal of mathematically-looking composition rules. Combined with a lack of knowledge of how his music actually sounded in 17th and 18th century (especially, whether the performer was allowed to decorate the sparse and restrained score with constant embellishments and theatrical effects), Bach has now acquired an image of a somewhat dry (even if genius) composer, almost of a musical engineer who cold-bloodedly built his impressive musical structures.

This, in turn, was a great fit for the new tool of the analog synthesizer, with its image of mathematical perfection and scientific groundings. Little did the listeners know how difficult it was in reality for the author of *Switched-On Bach* to keep the unreliable machine in tune. According to historical accounts, the physicality of the device, with all its analog circuitry, stood in direct contrast to the promises of mathematical purity, requiring the user to frequently readjust the tuning, and sometimes even to literally resort to the use of a hammer to make the machine do its work.[13]

In my view, this is a beautifully complete metaphor of mathematical music. For all its imagined idealist purity, any real-world realization of sound is still tied to materiality. And this materiality is, more often than not, messy, intertwined, and noisy. With these words, I would like to end my overview of mathematically created music, and turn to the conclusion, in which I discuss possible pathways of further inquiry, all leading away in different directions from the core topic of this book.

Notes

1 *How AI Could Compose a Personalized Soundtrack to Your Life.* TED, 2018. https://youtu.be/wYb3Wimn01s

2 Ellon D. Carpenter, "The Contributions of Taneev, Catoire, Conus, Garbuzov, Mazel, and Tiulin," in *Russian Theoretical Thought in Music*, ed. Gordon D. McQuere (Ann Arbor, MI: UMI Research Press, 1983), 294; Ellon D. Carpenter, "Russian Music Theory. A Conspectus," in *Russian Theoretical Thought in Music*, ed. Gordon D. McQuere (Ann Arbor, MI: UMI Research Press, 1983), 46.

3 Ellon D. Carpenter, "The Contributions of Taneev, Catoire, Conus, Garbuzov, Mazel, and Tiulin," in *Russian Theoretical Thought in Music*, ed. Gordon D. McQuere (Ann Arbor, MI: UMI Research Press, 1983), 303.

4 Frank Fischer, Marina Akimova, and Boris Orekhov, "Data-Driven Formalism," *Journal of Literary Theory* 13, no. 1 (2019): 1–12.

5 Olga Panteleeva, "How Soviet Musicology Became Marxist," *The Slavonic and East European Review* 97, no. 1 (2019): 99–100; Abraham Moles, *Information Theory and Esthetic Perception* (Urbana: University of Illinois Press, 1966).

6 In 1970, the famous West German avant-garde composer Karlheinz Stockhausen publicly claimed that "Electronic music began in Cologne in 1952-3," adding with a tone of indignation that "intellectual property is being stolen [by Americans], its source concealed and ultimately forgotten in order to suppress any historical sense, whereupon they claim the ideas as their own national product." Karlheinz Stockhausen, "The Origins of Electronic Music," *The Musical Times* 112, no. 1541 (1971): 649–650.

7 Karlheinz Stockhausen, "... How Time Passes ...," *Die Reihe [English Edition]*, Musical craftsmanship (1959): 10–40.

8 For an example of Schillinger's introductory exercises, see the illustration based on a drawing by Schillinger's pupil Lawrence Berk in the chapter on developments after 1900. Stockhausen includes on p. 17 a drawing showing different divisions of a time period stacked together which is very similar to the illustration on p. 644 in Schillinger's book *The Mathematical Basis of the Arts* (New York: Philosophical Library, 1948). Not being necessarily a proof of Stockhausen's knowledge of Schillinger's publication, this is still a sign that they both could have borrowed from the same earlier source of musical-mathematical thinking.

9 Already in 1931, an overview of the then recent technological developments in music was provided by Schillinger in a published article: Joseph Schillinger, "Electricity, a Musical Liberator," *Modern Music* 8 (March–April) (1931): 26–31.

10 The work done by the Russian-Soviet inventor Leon Theremin (his eponymous touch-free electronic instrument which he allegedly personally presented to Lenin) and by the German engineer Friedrich Trautwein (his so-called "Volkstrautonium" electronic instrument from 1933) is exemplary in this case.

11 Trevor J. Pinch and Frank Trocco, *Analog Days. The Invention and Impact of the Moog Synthesizer* (Cambridge, MA: Harvard University Press, 2004).

12 From all the synthesizers that decidedly avoided using the traditional piano keyboard layout as an input device, those by Don Buchla, a contemporary of Bob Moog, are among the most famous.

13 Amanda Sewell, "Switched-On Bach and Undesired Fame (1968–1969)," in *Wendy Carlos. A Biography* (Oxford: Oxford University Press, 2020).

CONCLUSION

Computer music has long been imagined as the outcome of the fact that computers existed. No doubt, the high operating speed and the information storage capacities of these calculating machines are prerequisites for the efficient execution of a music-creating program. They are indispensable if music needs to be generated automatically and on the fly. But musicians and theoreticians have long experimented with formal, rigid prescriptions on how to write acceptable music without the need, for the operator, to have a proper understanding of its rules and traditions. Sometimes, these prescriptions were so clear-cut and unambiguous that they could be turned into simple mechanisms that successfully imitated certain parts of the thinking process of a composer. By the early 19th century, a machine could be made that not only created randomized combinations of interchangeable snippets of music, but also actually played them through a built-in organ. This apparatus, the *Componium*, fascinated the public of its time. Since then, waves upon waves of technical and industrial revolutions have produced digital tools that are more reliable, fast, and capable of dealing with vastly bigger amounts of information than the simple pinned wheels which held the *Componium*'s musical material.

Now, data-gathering and personal profiling on the internet might for the first time give users of algorithms the possibility of mass-producing individualized music, not merely of works that are optimized for a large, anonymous group. Discussing the question whether this development should be seen as potentially harmful with regard to the manipulative and

DOI: 10.4324/9781003229254-17

ideological functions of mass culture was not the focus of this text. I expect studies of culture to present a political critique of such a scenario. Instead, in this book, I concentrated on showing, by relying on past examples, what is actually likely to be accepted by tomorrow's producers and audiences.

Which tools will be available tomorrow depends on the research that is ongoing today. Several technology corporations that have an interest in the development of artificial intelligence tools have been sponsoring basic research into music AI in the recent years, in addition to the university-based academic work. The results of this research are beginning to become accessible to the average user, and it can be anticipated that techniques stemming from these experiments will one day become integrated into the production of music.

But why do technology giants spend money on music AI research? After all, the sale of popular music is mostly not the central area of their business model. There are several answers to this question. First, there is prestige. Having researchers participate in academic conferences and publications who are associated not with a university, but with a company, is a way to promote this company's name and image. Second, theoretical work on AI in music might in principle also become useful for other AI-based tools in the context of a search engine or a social network.

In my view, however, a different, third cause is the most important one: A working music AI system, one that is broadly accepted as a meaningful solution by creators and audiences, is a proof that this technology does work as expected, and that it is not being seen by the majority as a harmful development. In effect, the future of music AI is therefore also inseparable from the agendas of those actors who are interested in the more general applications of this technology.

Now, let's look at some possible future implications of the topics discussed in this book.

There are several. In a previous chapter, I have written about the interesting capability of machine learning systems to imitate the results of rule-based composition without relying on an explicitly stated rulebook.[1] Such an imitation is certainly of interest to people wishing their generated music to be accepted by listeners as similar to "handmade" examples, and therefore as having a value comparable to human music. But could it be that the imitation of rule-based outcomes is not enough? After all, aesthetic effect is often achieved by subverting a well-established rule at a special point in the composition, after the general framework of underlying rules has been made evident through several predictable passages. A musician can, for example, use a chord they know does not fit the key of the piece, but which they nevertheless employ to express the idea of surprise, or to draw attention to the lyrics that occur at this point.

Human composers and musicians almost always rely on both their intuition derived from previous examples *and* some explicitly stated theory. In its reliance on broad tendencies derived from the training data, machine learning is in my view trying to replicate the intuition-based work of human composers. But what about the rules? Even the amateur musician who does not read musical notation, or the average listener, mostly know *something* about explicitly stated theory like, for example, the formal division of a song into parts (intro, chorus, verse, and bridge).

Having this knowledge, they are also able to create, and to understand, situations where rules have been subverted. And this is exactly the part that is probably not being imitated enough by musical machine learning systems. Therefore, I think that it can be beneficial to imitate more not just the regularities found in the data, but also the subversion of these regularities. Decisions on where and how this subversion should take place could also be derived from the data. With the right input and type of model, this task should not be beyond existing machine learning capabilities.

Another area of musical machine learning research that is in my view both fruitful and within the technical possibilities of today is the imitation of physical and mental boundaries of human composers and performers. Humans have limits, and the kind of music they make is conditioned by them. Anyone interested in creating human-like music through technical means needs to adhere to these limits, or the result will be much less acceptable as an imitation of a work made by a real person.

Some of these limits are very obvious. They are easy to describe mathematically, and to operationalize as a part of a program. Some are hidden, and need to be teased out through tests and psycho-physiological studies.

The easy ones are, for example, the following:

- A piano score should not contain notes that cannot be reached by an average human pianist (two hands, five fingers, each hand spanning not much more than an octave).
- With singing and with wind instruments, a score should not contain notes that are impossibly short or, conversely, so long that a human musician would run out of breath.
- The instrument, and the way a human musician uses it, limit the possibilities, too: a drummer, for example, normally has arms, feet and the instruments positioned in a specific way, making certain combinations of sounds unlikely or even impossible.

The more hidden limits on speed and complexity of performance also have their boundaries in the mental and physical capabilities of humans.

Again, everyone is free to ignore these limits for artistic effect or experimentation. But if the goal is to imitate human music, following these constraints is unavoidable. The reason is that the listener is used to hearing their *results*. Not taking them into account will make it less probable that the listener will feel that the output is similar to something produced by a person.

It is conceivable that some of these constraints might be efficiently imitated by manually coded rules. Previously, music programmers had notable success with such approaches. In the 1980s, a music-generating program created believable imitations of banjo playing by following typical picking patterns and by paying attention to the physical limitations of a human hand.[2] But, with today's technology and data, it is also conceivable that one can find a way to make a machine learning system *learn about* these constraints directly from data on human composition and performance.

As with any idea rooted in existing work and pointing into the future, I am sure these recommendations have already been considered and even used somewhere and in some form. The study of the history of science and technology makes it very obvious that there is always an earlier example for everything, yet it was not deemed important, convincing, or practical during its time. This is also in line with the underlying idea of this book that the existing state of music AI is the outcome of a blend of older developments. But I do believe that the ideas outlined above are especially powerful in today's specific situation.

Imagining the future of automated composition, one also cannot avoid contemplating some of the more negative trends in commercial and political online culture. Here, I will only briefly sketch out one speculative scenario for an undesirable development involving automated generation of music and functional sound, thus inviting the interested machine learning researcher or entrepreneur to also consider broader societal implications of their music-related work.

Under this imagined scenario, live information is collected on the psychological state of individual listeners, and used in advertising and propaganda campaigns.[3] The collection of this data would happen primarily, or officially, with the goal of providing "fitting" music, created for the mood and the circumstances of the current moment for each listener. Data could be provided explicitly by the listener, or deduced from other data such as the facial expression in a recent selfie, the tone of the voice, or even bodily indicators coming from a wearable health device. In return, the system would provide some music or sound which allegedly helps the listener "adjust" the mood and productivity to the desired level.

The motivation for the listener to use such a system would come not just from the wish to receive an artificial dose of happiness, calm, or energy, but also to feel individual and unique. A system which needs all that data, and all that extra computation, to create a very special musical product just for one person's one moment surely would not exist if this person were not so deeply original and extraordinary! Or would it?

The biggest problem, however, is that data travels in unpredictable ways. A music company could choose to share live psychological data with an advertising company (or to become such a company), which could then try to exploit the state of each listener with fitting messages—commercial, or even political. In the worst scenario, this could make large groups of people more vulnerable to manipulation (although a not really working system which is only claimed by its owners to be able to manipulate the public is also perfectly imaginable).[4]

My research into the history of mathematical methods in music first started almost a decade ago, when I began to make plans for my PhD dissertation in which I analyzed early music-generating video games and the culture and technology that prepared the ground for their emergence. Seen from today's perspective, that time looks like an almost completely different world. I wrote this present book in 2020–2021, during the coronavirus pandemic that turned the world of music, like everything else, on its head: concerts became impossible, endangering the livelihood of many practicing musicians, and a deep cultural turn towards even more digitization of work and leisure was gaining traction. The outcome of this transformation is likely to accelerate the rate at which technological, mathematical, and data-based tools are adopted in music, both for commercial and for artistic goals. With this book, I now wanted to equip the reader with a handy filter through which the discussions that are going to accompany future changes in music technology should ideally become better understandable and, hopefully, easier to navigate.

Having gained a bird's eye view of mathematical methods of creating music by reading this volume, you, the reader, can now continue in many directions. In addition to the literature cited in the respective chapters, plenty of material is available for further study.

Here, I have focused especially on popular and applied music, both historical (like the Quadrille Melodist from the 19th century) and contemporary (like the randomized relaxation soundscapes offered by a start-up from Berlin). But there is a whole world of very interesting experiments in art music, and of artistic and academic commentary on these experiments. For this topic, excellent overview literature exists, with which one can embark on a life-long journey of listening and, possibly, even creation.[5]

Those interested in learning to apply programming and machine learning in a practical musical scenario will find many helpful resources in book publications and on the internet.[6]

Good resources are available for readers interested in a critical analysis of modern technology and of discourses that surround it.[7]

Finally, I would like point out that, now that you have joined me on this first, linear journey through the history of mathematical music, you are well equipped to explore in more detail all the loops, breaks, dead ends, and rediscoveries that happened in the past and continue to repeat themselves today. I chose for this book, and especially for its first, historical, part, the form of a linear narrative that goes from the earliest examples to the most recent. In my view, it is the best possible way to introduce someone to a topic in a way that facilitates understanding and overview. But a linear narrative is, of course, not the only possible way to describe what happened. The exploration of mathematical music is an open-ended process. Thank you for joining in.

Notes

1 See the chapter "Putting music AI in perspective."
2 *IMG/1 Incidental Music Generator—a Conversation between Eamonn Bell and Nikita Braguinski*, 2021. https://youtu.be/4_3ehJs-uHo. See also the description of Peter Langston's work in the chapter on developments since 1950.
3 For an example of discussions surrounding the use of emotion-detecting technology in the context of music, see Mark Savage, "Spotify Wants to Suggest Songs Based on Your Emotions," *BBC News*, January 28, 2021. https://www.bbc.com/news/entertainment-arts-55839655
4 For a deeper perspective on potentially harmful uses of similar technologies, and on the sometimes distorted presentation of the abilities of such technologies, see: Fenwick McKelvey, "The Other Cambridge Analytics. Early 'Artificial Intelligence' in American Political Science," in *The Cultural Life of Machine Learning: An Incursion into Critical AI Studies*, ed. Jonathan Roberge and Michael Castelle (Cham: Palgrave Macmillan, 2021), 117–142; Robert Prey, "Nothing Personal. Algorithmic Individuation on Music Streaming Platforms," *Media, Culture & Society* 40, no. 7 (2018): 1086–1100; Bernd Bösel, "Affective Computing," in *Mensch-Maschine-Interaktion. Handbuch zu Geschichte – Kultur – Ethik*, ed. Kevin Liggieri and Oliver Müller (Stuttgart: J.B. Metzler, 2019), 223–225 (in German).
5 Nick Collins and Julio d'Escriván, *The Cambridge Companion to Electronic Music* (Cambridge: Cambridge University Press, 2017). See also the back issues of *Leonardo Music Journal*, a publication that, for 30 years, has collected many artist statements and analytic ideas in this area: https://www.leonardo.info/leonardo-music-journal

6 See, for example: Gerhard Nierhaus, *Algorithmic Composition. Paradigms of Automated Music Generation* (Wien: Springer, 2009); Jean-Pierre Briot, Gaëtan Hadjeres, and François-David Pachet, *Deep Learning Techniques for Music Generation*, Computational Synthesis and Creative Systems (Springer International Publishing, 2020); Google Magenta, https://magenta.tensorflow.org/

7 Mark Coeckelbergh, *AI Ethics* (Cambridge, MA: MIT Press, 2020); Robert Seyfert and Jonathan Roberge, eds., *Algorithmic Cultures. Essays on Meaning, Performance and New Technologies* (London: Routledge, 2016); Jonathan Roberge and Michael Castelle, eds., *The Cultural Life of Machine Learning: An Incursion into Critical AI Studies* (Cham: Palgrave Macmillan, 2021).

GLOSSARY

This glossary contains brief, non-specialist explanations of some of the terms used in this book. It focuses on the specific meanings that they have in the context of mathematically created music.

acoustics - The study of the physical properties of sound. At times, this term was also used in relation to a combined study of the sound's physics and perception.

aesthetics - The discipline studying notions and perceptions related to art and beauty.

aleatoric music - A branch of primarily 20th-century experimental music employing random and unpredictable elements.

algorithm - A scheme containing instructions (such as "multiply X with 2") and rules for navigating it (such as "if X is greater than 5, continue with the instruction ABC").

analog synthesis - Electronic music technology, originally used in the second half of the 20th century. Its most popular form, the subtractive synthesis, consists of a sound source (oscillator or noise generator), and of the means to influence both the parameters of the source and the changes applied to the sound after being generated (filter, loudness contour, etc.).

arithmetic - The well-known part of mathematics concerned with such operations as addition or multiplication.

artificial intelligence (AI) - A term which has been employed and understood in many different ways since its first modern use in the middle of the 20th century. Depending on the source, its exact meaning may vary. Today, it is commonly used to refer to technical solutions based on machine learning.

avant-garde - An arts movement aiming to overthrow past traditions and techniques. Some of the notable avant-gardes of the 20th century were the movements of the 1920s, including Russian musicians and theorists, and the post-World War II musical avant-garde in Western countries.

chord - In Western musical tradition, a combination of mostly at least three notes, defined by specific intervals (relations) between the notes.

combinatorics - A branch of mathematics concerned with counting and listing possible groupings of individual elements.

consonance - In Western music theory, the quality of several tones to sound "good" together.

cybernetics - A discipline that, in the early days of computer technology, began to study different areas (technical, biological, etc.) under a shared systems-oriented approach.

dissonance - The opposite of consonance.

frequency - In its basic sense, a number showing how often something happens in a period of time (such as 440 vibrations per second). Real-world musical sounds normally consist of a combination of many different regularly repeating vibrations, each with its own frequency, as well as of some non-regular, noisy components.

harmonics - Individual parts in a complex sound whose frequencies are multiples of the frequency of its slowest constituent part. For example, if the slowest part has the frequency of 100 vibrations per second, harmonics may have frequencies of 200, 300, etc.

harmony - The area of Western music theory concerned with rules for the construction of chords and for combining them into meaningful sequences.

just intonation - A tradition inside music theory postulating that notes whose frequencies are related by a small-integer ratio have special aesthetic qualities, often referring to notes that form a musical scale.

machine learning - A programming approach to the problem of segmentation and generation of information in which existing data is used to semi-automatically "train" the system.

microtone - An interval that is narrower than the smallest one traditionally used in Western music.

model - In machine learning, the technically and mathematically defined structure capable of being "trained" and of being used to segment or to generate information.

modernism – A current in arts history aiming for a renewal of cultural practice, specifically the movements that were active at the beginning of the 20th century.

neural network – In machine learning, a structure consisting of interconnected neurons (simple calculating units), capable of being "trained."

octave – A musical interval in which the frequency of the higher tone is twice that of the lower one.

permutation – In combinatorics, an operation by which elements are put into a certain order. In relation to music, this word is mostly used to denote the rearrangement of notes or musical snippets in a new way, without leaving out or repeating any of the original elements.

pitch – In its more precise sense, the perceptual quality of frequency, which is the underlying physical phenomenon of pitch (probably, together with tone color and other parameters). In the more general sense, however, this word is often used to simply refer to tone height (this is also how I use this word in this book).

positivism – In the context of historical music aesthetics research, an intellectual current focusing especially on questions and methods that can be used in practical experiments, as opposed to those more suitable for speculation or subjective introspection.

procedure – In the context of formal rule-based composition and computer music, a clearly defined course of action by which a certain outcome can be reached (such as constructing a melody or rhythm).

pseudorandomness – A quality, of a row of numbers, of appearing seemingly random to a human observer, or of being statistically random for a certain use, despite the fact that they are actually created by following a mathematical procedure. Pseudorandom numbers are commonly used in computer applications to compensate for a lack of a dedicated electronic part capable of generating true randomness.

randomness – A quality, of numbers in a row, of being genuinely unrelated to each other, so that each number does not influence the probability of any other number occurring. This results in an unpredictable and unstructured stream of numbers that can be used as control parameters in various applications, including the generation of music.

ratio – A description of relations. A ratio of 1:2 means that one thing is two times bigger, or longer, or louder, and so on, than another.

reinforcement learning – A branch of machine learning concerned with the simulation of the behavior of a simple robot (called "agent") acting inside a controlled environment with the goal of letting the robot learn the most efficient course of action through trial and error.

rhythm - A musical concept relating to how musical events such as audible notes are structured in time. Often, this word implies that such events are repeated regularly, forming a discernible pattern. In a second sense, rhythm describes the temporal structure of music, regardless whether the pattern is regular or not.

tone - The audible phenomenon behind the cultural construction called "note."

tuning - In music theory, the system of mathematically defined intervals between notes forming a scale.

unsupervised learning - In machine learning, a form of "training" in which the correct result is not provided at the beginning.

waveform - In its most general sense, the shape that a vibrating part (such as the string of a guitar) would have painted onto a strip of paper if the paper were constantly moving, creating a visual representation of this vibration. More specifically, this word is often used to describe short repeating patterns of vibration, especially in analog synthesis.

BIBLIOGRAPHY

Anderson, Robert, Arturo Chamorro, Ellen Hickmann, Anne Kilmer, Gerhard Kubik, Thomas Turino, Vincent Megaw, and Alan R. Thrasher. "Archaeology of Instruments." *Grove Music Online*, 2020. https://doi.org/10.1093/gmo/9781561592630.article.L2293842.

Avdeeff, Melissa. "Artificial Intelligence & Popular Music: SKYGGE, Flow Machines, and the Audio Uncanny Valley." *Arts* 8, no. 4 (2019): 130. https://doi.org/10.3390/arts8040130.

Barbera, André. "Pythagoras." *Grove Music Online*, 2001. https://doi.org/10.1093/gmo/9781561592630.article.22603.

Birtchnell, Thomas. "Listening without Ears: Artificial Intelligence in Audio Mastering." *Big Data & Society* 5, no. 2 (July 1, 2018): 1–16. https://doi.org/10.1177/2053951718808553.

Bojarinov, Denis. "Dmitry Evgrafov: 'Iz sta melodij, kotorye mne sdelala nejronnaja set', ja polovinu vykinul." *Colta.ru*, August 5, 2020. https://www.colta.ru/articles/music_modern/25073-dmitriy-evgrafov-intervyu-albom-surrender.

Bonner, Anthony. *The Art and Logic of Ramon Llull. A User's Guide*. Leiden; Boston: Brill, 2007.

Bösel, Bernd. "Affective Computing." In *Mensch-Maschine-Interaktion. Handbuch zu Geschichte – Kultur – Ethik*, edited by Kevin Liggieri and Oliver Müller, 223–225. Stuttgart: J.B. Metzler, 2019. https://doi.org/10.1007/978-3-476-05604-7_30.

Bower, Calvin M. "The Transmission of Ancient Music Theory into the Middle Ages." In *The Cambridge History of Western Music Theory*, edited by Thomas Christensen, 136–167. Cambridge: Cambridge University Press, 2006.

Braguinski, Nikita. "'428 Millions of Quadrilles for 5s. 6d.': John Clinton's Combinatorial Music Machine." *19th-Century Music* 23, no. 2 (Fall 2019): 86–98. https://doi.org/10.1525/ncm.2019.43.2.86.

―――. "Die Systeme der reinen Stimmung von August Eduard Grell und ihr geistesgeschichtlicher Kontext." In *Jahrbuch 2011 des Staatlichen Instituts für Musikforschung Preußischer Kulturbesitz*, 75–104. Mainz: Schott, 2011.

―――. "Musofun. Joseph Schillinger's Musical Game between American Music, the Soviet Avant-Garde, and Combinatorics." *American Music* 38, no. 1 (Spring 2020): 55–77.

―――. *RANDOM. Die Archäologie der elektronischen Spielzeugklänge.* Computerarchäologie 3. Bochum: Projekt Verlag, 2018.

Bréard, Andrea. "China." In *Combinatorics: Ancient and Modern*, edited by Robin Wilson and John J. Watkins, 65–82. Oxford: Oxford University Press, 2013.

Bretanickaja, Alla, ed. *Dve žizni Iosifa Šillingera. Žizn' pervaja. Rossija. Žizn' vtoraja. Amerika.* Moskva: Moskovskaja konservatorija, 2015.

Briot, Jean-Pierre, Gaëtan Hadjeres, and François-David Pachet. *Deep Learning Techniques for Music Generation.* Computational Synthesis and Creative Systems. Springer International Publishing, 2020. https://www.springer.com/gp/book/9783319701622.

Bröcker, Marianne. "Äolsharfe." MGG Online, 1994. https://www.mgg-online.com/mgg/stable/11719.

Brodsky, Warren. "Joseph Schillinger (1895–1943): Music Science Promethean." *American Music* 21, no. 1 (2003): 45–73. https://doi.org/10.2307/3250556.

Busoni, Ferruccio. *Entwurf einer neuen Aesthetik der Tonkunst.* Berlin: Berliner Musikalien Druckerei, 1907. https://busoni-nachlass.org/edition/essays/E010004/D0200001.

Carpenter, Ellon D. "Russian Music Theory. A Conspectus." In *Russian Theoretical Thought in Music*, edited by Gordon D. McQuere, 83–108. Ann Arbor, MI: UMI Research Press, 1983a.

―――. "The Contributions of Taneev, Catoire, Conus, Garbuzov, Mazel, and Tiulin." In *Russian Theoretical Thought in Music*, edited by Gordon D. McQuere, 253–378. Ann Arbor, MI: UMI Research Press, 1983b.

Carr, David. "Giving Viewers What They Want." *The New York Times*, February 24, 2013, sec. Business. https://www.nytimes.com/2013/02/25/business/media/for-house-of-cards-using-big-data-to-guarantee-its-popularity.html.

Chen, Ching-Wei, and Murali Vidhya. "Machine Learning and Big Data for Music Discovery at Spotify. Presentation." Galvanize NYC. March 9, 2017.. https://www.slideshare.net/cweichen/machine-learning-and-big-data-for-music-discovery-at-spotify?qid=b7ca7727-0a01-4441-9107-14410ccf0e7d.

Coeckelbergh, Mark. *AI Ethics.* Cambridge, MA: MIT Press, 2020.

Collins, Nick. "Origins of Algorithmic Thinking in Music." In *The Oxford Handbook of Algorithmic Music*, edited by Alex McLean and Roger T. Dean, 67–78. Oxford: Oxford University Press, 2018.

Collins, Nick, and Julio d'Escriván. *The Cambridge Companion to Electronic Music.* Cambridge: Cambridge University Press, 2017.

Dan, Joseph. *Kabbalah. A Very Short Introduction.* Oxford: Oxford University Press, 2006.

Deutsch, Diana. "Grouping Mechanisms in Music." In id. *The Psychology of Music*, 299–348. San Diego, CA; London: Academic Press, 1999.

Dhariwal, Prafulla, Heewoo Jun, Christine Payne, Jong Wook Kim, Alec Radford, and Ilya Sutskever. "Jukebox: A Generative Model for Music." *ArXiv:2005.00341 [Cs, Eess, Stat]*, April 30, 2020. http://arxiv.org/abs/2005.00341.

Di Nunzio, Alex. "Push Button Bertha." *Musica Informatica*, 2013. http://www.musicainformatica.org/topics/push-button-bertha.php.

Dick, Stephanie. "Of Models and Machines. Implementing Bounded Rationality." *Isis* 106, no. 3 (September 2015): 623–634. https://doi.org/10.1086/683527.

Doornbusch, Paul. *The Music of CSIRAC. Australia's First Computer Music.* Australia: Common Ground Publishing, 2005.

Dostrovsky, Sigalia, Murray Campbell, James F. Bell, and C. Truesdell. "Physics of Music." *Grove Music Online*, 2001. https://doi.org/10.1093/gmo/9781561592630. article.43400.

Dostrovsky, Sigalia, and John T. Cannon. "Entstehung der musikalischen Akustik." In *Hören, Messen und Rechnen in der frühen Neuzeit*, edited by Frieder Zaminer, 7–79. Darmstadt: Wissenschaftliche Buchgesellschaft, 1987.

Ebbeke, Klaus. "Aleatorik." *MGG Online*, 1994. https://www.mgg-online.com/mgg/stable/16100.

Eriksson, Maria, Rasmus Fleischer, Anna Johansson, Pelle Snickars, and Patrick Vonderau. *Spotify Teardown: Inside the Black Box of Streaming Music.* Cambridge, MA: The MIT Press, 2018.

Ferreira, Manuel Pedro. "Proportions in Ancient and Medieval Music." In *Mathematics and Music. A Diderot Mathematical Forum*, edited by Gerard Assayag, Hans Georg Feichtinger, and Jose Francisco Rodrigues, 1–26. Berlin: Springer, 2002.

Finkel, Irving L, ed. *Ancient Board Games in Perspective. Papers from the 1990 British Museum Colloquium.* London: British Museum Press, 2007.

"First Recording of Computer-Generated Music – Created by Alan Turing – Restored." *The Guardian*, September 26, 2016. https://www.theguardian.com/science/2016/sep/26/first-recording-computer-generated-music-created-alan-turing-restored-enigma-code.

Fischer, Frank, Marina Akimova, and Boris Orekhov. "Data-Driven Formalism." *Journal of Literary Theory* 13, no. 1 (2019): 1–12. https://doi.org/10.1515/jlt-2019-0001.

Gioti, Artemi-Maria. "From Artificial to Extended Intelligence in Music Composition." *Organised Sound* 25, no. 1 (2020): 25–32. https://doi.org/10.1017/S1355771819000438.

Gray, J. "Computational Imaginaries. Some Further Remarks on Leibniz, Llull, and Rethinking the History of Calculating Machines." In *Dia-Logos: Ramon Llull's Method of Thought and Artistic Practice*, edited by Amador Vega, Peter Weibel, and Siegfried Zielinski, 293–300. Minneapolis: University of Minnesota Press, 2018.

Grell, Eduard. *Aufsätze und Gutachten über Musik.* Edited by Heinrich Bellermann. Berlin: Springer, 1887.

Guerrieri, Matthew. "'Automation Divine'. Early Computer Music and the Selling of the Cold War." *NewMusicBox*, October 10, 2018. https://nmbx.newmusicusa.org/automation-divine-early-computer-music-and-the-selling-of-the-cold-war/.

Hayes, Tyler. "The Science behind Endel's AI-Powered Soundscapes." *Amazon Science*, November 25, 2020. https://www.amazon.science/latest-news/the-science-behind-endels-ai-powered-soundscapes.

Hazell, Ed. *Berklee: The First Fifty Years.* Edited by Lee Eliot Berk. Berklee Press Publications, 1995.

Helmholtz, Hermann von. *On the Sensations of Tone as a Physiological Basis for the Theory of Music.* Second English Edition. London: Longmans, Green, 1885.

Hiller, Lejaren, and Leonard Isaacson. *Experimental Music. Composition with an Electric Computer.* New York: McGraw-Hill, 1959.

Hoffmann, Peter. "Xenakis, Iannis." *Grove Music Online,* 2016. https://doi.org/10.1093/gmo/9781561592630.article.30654.

How AI Could Compose a Personalized Soundtrack to Your Life. Pierre Barreau. TED, 2018. https://youtu.be/wYb3Wimn01s.

Huang, Cheng-Zhi Anna, Hendrik Vincent Koops, Ed Newton-Rex, Monica Dinculescu, and Carrie J. Cai. "AI Song Contest: Human-AI Co-Creation in Songwriting." *Magenta,* October 13, 2020. https://magenta.tensorflow.org/aisongcontest.

Hyde, Ralph. "Myrioramas, Endless Landscapes. The Story of a Craze." *Print Quarterly* 21, no. 4 (2004): 403–421.

IMG/1 Incidental Music Generator—A Conversation between Eamonn Bell and Nikita Braguinski, 2021. https://youtu.be/4_3ehJs-uHo.

Katz, Victor J. "Jewish Combinatorics." In *Combinatorics: Ancient and Modern,* edited by Robin Wilson and John J. Watkins, 109–122. Oxford: Oxford University Press, 2013.

Keislar, Douglas. "A Historical View of Computer Music Technology." In *The Oxford Handbook of Computer Music,* edited by Roger T. Dean, 11–43. Oxford: Oxford University Press, 2009.

Kelih, Emmerich. "Quantitative Verfahren in der russischen Literaturwissenschaft der 1920er und 1930er Jahre." In *Quantitative Ansätze in den Literatur- und Geisteswissenschaften,* 269–288. De Gruyter, 2018. https://doi.org/10.1515/9783110523300-012.

Kelleher, John D. *Deep Learning.* Cambridge, MA: MIT Press, 2019.

Keller, Damián, and Brian Ferneyhough. "Analysis by Modeling. Xenakis's ST/10–1 080262." *Journal of New Music Research* 33, no. 2 (2004): 161–171. https://doi.org/10.1080/0929821042000310630.

Kircher, Athanasius. *Musurgia Universalis Sive Ars Magna Consoni et Dissoni in X. Libros Digesta.* Rome: Corbelletti; Grignani, 1650.

Kirnberger, Johann Philipp. *Der allezeit fertige Polonoisen- und Menuettenkomponist.* Berlin, 1757. http://mdz-nbn-resolving.de/urn:nbn:de:bvb:12-bsb10527349-7.

Klein, Martin L. "Syncopation by Automation." *Radio-Electronics* 28, no. 6 (June 1957): 36–38.

Klotz, Sebastian. *Kombinatorik und die Verbindungskünste der Zeichen in der Musik zwischen 1630 und 1780.* Berlin: Akademie, 2006.

Knobloch, Eberhard. "The Sounding Algebra. Relations Between Combinatorics and Music from Mersenne to Euler." In *Mathematics and Music. A Diderot Mathematical Forum,* edited by Gerard Assayag, Hans Georg Feichtinger, and Jose Francisco Rodrigues, 27–48. Berlin: Springer, 2002.

Knuth, Donald E. "Two Thousand Years of Combinatorics." In *Combinatorics: Ancient and Modern,* edited by Robin Wilson and John J. Watkins, 3–37. Oxford: Oxford University Press, 2013.

Koetsier, Teun. "The Art of Ramon Llull (1232–1350). From Theology to Mathematics." *Studies in Logic, Grammar and Rhetoric*, 44, no. 57 (2016): 55–80.

Kursell, Julia. *Epistemologie des Hörens: Helmholtz' physiologische Grundlegung der Musiktheorie.* Paderborn: Fink, 2018.

———. "Hermann von Helmholtz und Carl Stumpf über Konsonanz und Dissonanz." *Berichte zur Wissenschaftsgeschichte* 31, no. 2 (2008): 130–143. https://doi.org/10.1002/bewi.200801314.

Langston, Peter. "Six Techniques for Algorithmic Music Composition. A Paper for the 15th International Computer Music Conference (ICMC)." Columbus, Ohio, November 2–5, 1989. http://www.langston.com/Papers/amc.pdf.

Loy, Gareth. *Musimathics: The Mathematical Foundations of Music.* Cambridge; London: MIT Press, 2006.

Magnusson, Thor. *Sonic Writing. Technologies of Material, Symbolic, and Signal Inscriptions.* London: Bloomsbury Academic, 2019.

McKay, John Zachary. *Universal Music-Making: Athanasius Kircher and Musical Thought in the Seventeenth Century.* Doctoral Dissertation, Harvard University, 2013. http://nrs.harvard.edu/urn-3:HUL.InstRepos:10382782.

McKelvey, Fenwick. "The Other Cambridge Analytics. Early 'Artificial Intelligence' in American Political Science." In *The Cultural Life of Machine Learning: An Incursion into Critical AI Studies,* edited by Jonathan Roberge and Michael Castelle, 117–142. Cham: Palgrave Macmillan, 2021.

McLean, Alex, and Roger T. Dean, eds. *The Oxford Handbook of Algorithmic Music.* Oxford: Oxford University Press, 2018.

Mersenne, Marin. *Harmonie universelle, contenant la theorie et la pratique de la musique.* Paris: Sebastien Cramoisy, 1636.

Moles, Abraham. *Information Theory and Esthetic Perception.* Urbana: University of Illinois Press, 1966.

Nicholls, David. "Brave New Worlds: Experimentalism between the Wars." In *The Cambridge History of Twentieth-Century Music,* edited by Nicholas Cook and Anthony Pople, 210–227. Cambridge: Cambridge University Press, 2004.

Nierhaus, Gerhard. *Algorithmic Composition. Paradigms of Automated Music Generation.* Wien: Springer, 2009.

Nolan, Catherine. "Music Theory and Mathematics." In *The Cambridge History of Western Music Theory,* edited by Thomas Christensen, 272–304. Cambridge: Cambridge University Press, 2006.

Ong, Walter J. *Orality and Literacy. The Technologizing of the Word. With Additional Chapters by John Hartley.* London: Routledge, 2012.

Panteleeva, Olga. "How Soviet Musicology Became Marxist." *The Slavonic and East European Review* 97, no. 1 (2019): 73–109. https://doi.org/10.5699/slaveasteurorev2.97.1.0073.

Patteson, Thomas. *Instruments for New Music: Sound, Technology, and Modernism.* Oakland: University of California Press, 2016.

Pesic, Peter. *Music and the Making of Modern Science.* Cambridge, MA: The MIT Press, 2014.

Pinch, Trevor J., and Frank Trocco. *Analog Days. The Invention and Impact of the Moog Synthesizer.* Cambridge, MA: Harvard University Press, 2004.

Prey, Robert. "Nothing Personal. Algorithmic Individuation on Music Streaming Platforms." *Media, Culture & Society* 40, no. 7 (2018): 1086–1100. https://doi.org/10.1177/0163443717745147.

Riley, Terrance. "Composing for the Machine." *European Romantic Review* 20, no. 3 (2009): 367–379. https://doi.org/10.1080/10509580902986344.

Roberge, Jonathan, and Michael Castelle, eds. *The Cultural Life of Machine Learning: An Incursion into Critical AI Studies.* Cham: Palgrave Macmillan, 2021.

Sabaneev, Leonid L. "Zolotoe sečenie v prirode, v iskusstve i v žizni čeloveka (1959)." In *Vospominanija o Rossii.* Moscow: Klassika-XXI, 2004.

Salinas, Francisco de. *De musica libri septem.* Salmantica: Gastius, 1577.

Savage, Mark. "Spotify Wants to Suggest Songs Based on Your Emotions." *BBC News,* January 28, 2021. https://www.bbc.com/news/entertainment-arts-55839655.

Schillinger, Joseph. "Electricity, a Musical Liberator." *Modern Music* 8 (March–April) (1931): 26–31.

———. *The Mathematical Basis of the Arts.* New York: Philosophical Library, 1948.

Schuijer, Michiel. *Analyzing Atonal Music. Pitch-Set Theory and Its Context.* Rochester, NY: University of Rochester Press, 2008.

Seabrook, John. *The Song Machine: Inside the Hit Factory.* London: Vintage, 2016.

Sewell, Amanda. "Switched-On Bach and Undesired Fame (1968–1969)." In *Wendy Carlos. A Biography.* Oxford: Oxford University Press, 2020.

Seyfert, Robert, and Jonathan Roberge, eds. *Algorithmic Cultures. Essays on Meaning, Performance and New Technologies.* London: Routledge, 2016.

Shannon, Claude. "A Mathematical Theory of Communication." *The Bell System Technical Journal* 27, no. 3 (July 1948): 379–423. https://doi.org/10.1002/j.1538-7305.1948.tb01338.x.

Simon, Ian, Cheng-Zhi Anna Huang, Jesse Engel, Curtis Hawthorne, and Monica Dinculescu. "Generating Piano Music with Transformer." *Magenta,* September 16, 2019. https://magenta.tensorflow.org/piano-transformer.

Skinner, Rebecca E. "Artificial Intelligence." In *Debugging Game History. A Critical Lexicon,* edited by Henry Lowood and Raiford Guins, 29–36. Cambridge, MA: The MIT Press, 2016.

Skrynnikova, Anastasiya. "'My ne sozdajom muzyki': osnovatel' servisa generacii zvukovogo fona Endel o sdelke s Warner Brothers i rabote algoritma." *vc.ru,* April 10, 2019. https://vc.ru/future/63710-my-ne-sozdaem-muzyku-osnovatel-servisa-generacii-zvukovogo-fona-endel-o-sdelke-s-warner-music-i-rabote-algoritma.

Spotify Engineering. "For Your Ears Only: Personalizing Spotify Home with Machine Learning," January 16, 2020. https://engineering.atspotify.com/2020/01/16/for-your-ears-only-personalizing-spotify-home-with-machine-learning/.

Spotify for Developers. "Get Audio Features for a Track." Accessed January 6, 2021. https://developer.spotify.com/documentation/web-api/reference/tracks/get-audio-features/.

Steege, Benjamin. *Helmholtz and the Modern Listener.* Cambridge: Cambridge University Press, 2012.

Steinbeck, Wolfram. "Würfelmusik." *MGG Online,* 1998. https://www.mgg-online.com/mgg/stable/12552.

Sterne, Jonathan. "Out With the Trash. On the Future of New Media." In *Residual Media*, edited by Charles R. Acland, 16–31. Minneapolis, London: University of Minnesota Press, 2007.

Sterne, Jonathan, and Elena Razlogova. "Machine Learning in Context, or Learning from LANDR: Artificial Intelligence and the Platformization of Music Mastering." *Social Media + Society* April–June (2019): 1–18. https://doi.org/10.1177/2056305119847525.

Stockhausen, Karlheinz. "... How Time Passes..." *Die Reihe [English Edition]* Musical craftsmanship (1959): 10–40.

———. "The Origins of Electronic Music." *The Musical Times* 112, no. 1541 (1971): 649–650. https://doi.org/10.2307/957006.

Sutherland, Mark. "Songwriting: Why It Takes More than Two to Make a Hit Nowadays." *Music Week*, May 16, 2017. https://www.musicweek.com/publishing/read/songwriting-why-it-takes-more-than-two-to-make-a-hit-nowadays/068478.

Tiggelen, Philippe John van. *Componium: The Mechanical Musical Improvisor.* Louvain-la-Neuve: Institut supérieur d'archéologie et d'histoire de l'art, 1987.

Tommaso, Rocchi. "How Data Is Redefining the Role of A&R in the Music Industry Today." *blog.chartmetric.com*, September 21, 2020. https://blog.chartmetric.com/what-is-a-and-r-music-industry-today/.

Wicks, Robert. *European Aesthetics. A Critical Introduction from Kant to Derrida.* London: Oneworld, 2013.

Xenakis, Iannis. *Formalized Music. Thought and Mathematics in Composition.* Stuyvesant, NY: Pendragon, 1992.

Zielinski, Siegfried. *Deep Time of the Media. Toward an Archaeology of Hearing and Seeing by Technical Means.* Cambridge, MA: The MIT Press, 2006.

INDEX

Printed in the United States
by Baker & Taylor Publisher Services